Robotics Engineering Using LEGOs®

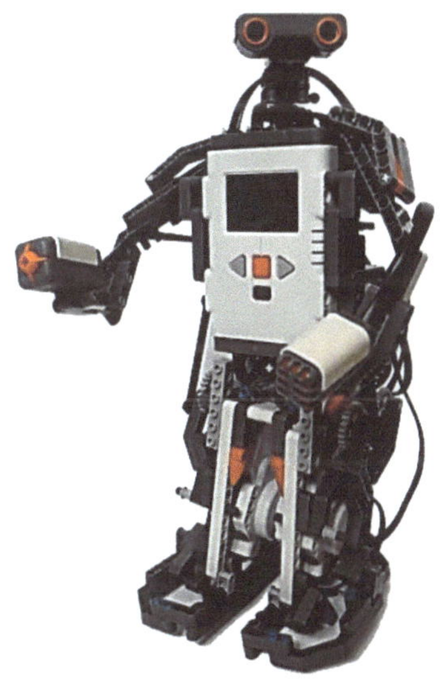

Ashley Rupprecht

Robotics Engineering Using LEGOs®

About the Author

Mrs. Ashley Rupprecht, often called by her students "The LEGO® Lady," is a robotics engineering teacher and LEGO® expert. She has been teaching physical science and computer technology courses to students all over Palm Beach County for the past six years. Ashley was homeschooled her entire life and has a Florida Teacher Certification and a B.A. degree in Education, *summa cum laude*, from Florida Atlantic University. She is the founder of MechaniKids™, an educational service which offers courses for students in S.T.E.M. areas using LEGOs®, which promote critical thinking and problem solving in an interactive, hands-on way. Ashley enjoys studying God's Word, worshipping at church, swimming, horseback riding, spending time with her family, and using LEGOs® to build new and productive robots. She lives in South Florida with her husband Jonathan, and their daughter, Anna. She can be contacted at ashley@mechanikids.com. Her website is www.mechanikids.com

When Mr. and Mrs. Rupprecht got married, they designed, made, and decorated their wedding cake, complete with an army of minifig cake decorators/construction workers, and a robotic crane.

Robotics Engineering Using LEGOs®

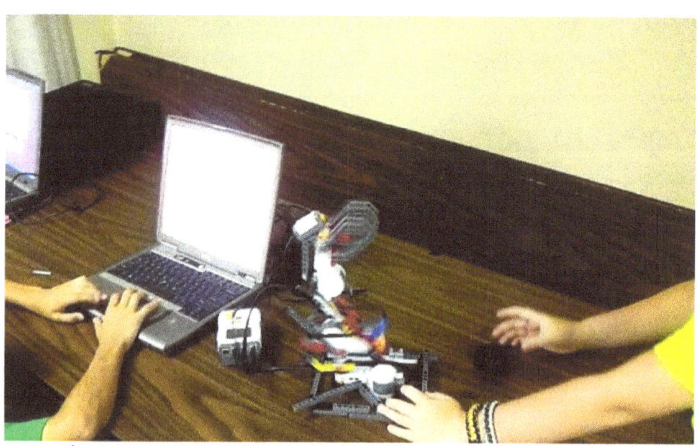

In this introductory robotics course, students will use their creativity and teamwork to design, create, and build full-fledged LEGO® robots that are programmed to complete tasks by using computers. Each class will promote experiences and learning in science, technology, engineering, and mathematics (S.T.E.M.) areas using LEGO® in an interactive, hands-on way.

Students will program the robots that they build using light, touch, ultrasonic, and sound sensors to perform mission tasks, and learn all about the science concepts and technology used in the field of robotics. While learning about engineering and robotics applications, the students will develop critical and higher-level thinking skills to solve problems by working cooperatively to plan, construct, and complete missions.

Students will learn how to do age-appropriate graphical computer programming using NXT-G while they expand their knowledge of robotic utilizations and task solving. This course is a fun way to learn robotics and critical thinking, and is a must for any aspiring engineer or young inventor!!

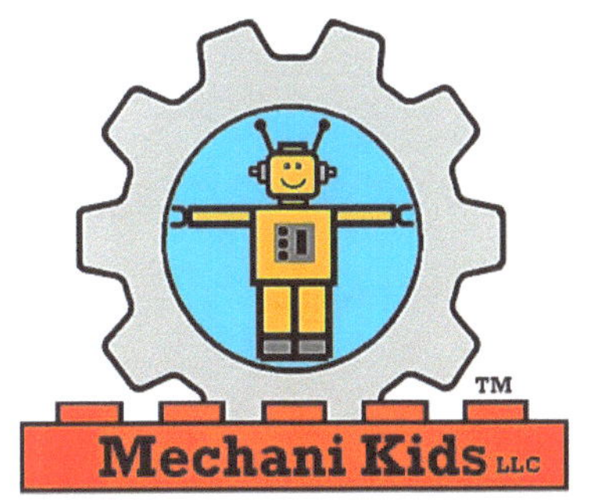

Table of Contents

UNIT 1
INTRODUCTION TO ROBOTICS

Photo: NASA

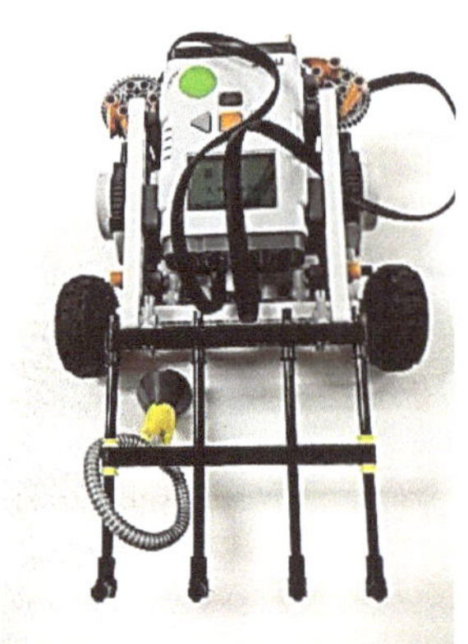

Photo: NASA

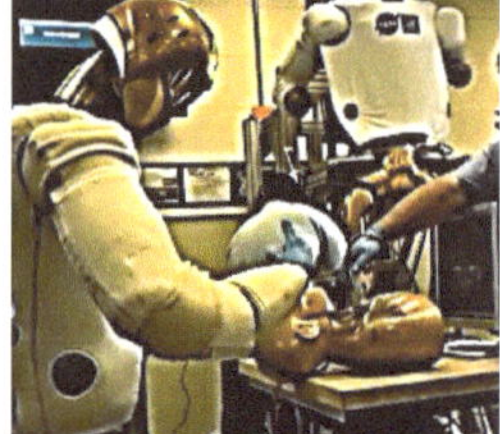
Photo: NASA

Welcome on an adventure to learn about robotics engineering!

Get your creative thinking caps on as we use LEGO® robots to learn more about this exciting and growing field of robotic science!

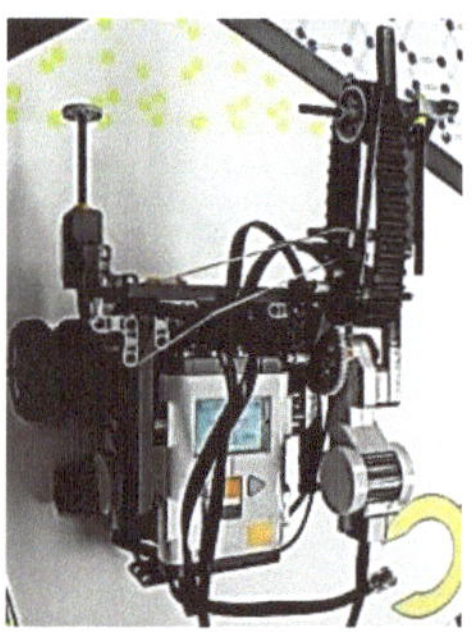

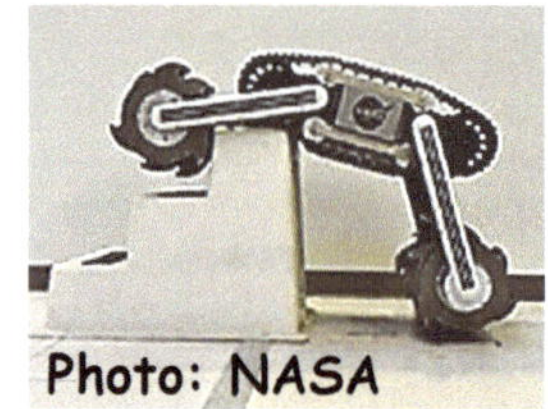
Photo: NASA

Photo: NASA

Photo: NASA

Photo: NASA

What Are Robots?

•What is a robot?

When you hear the word robot, a few famous movie robots may pop into your mind, such as R2D2 and C3PO from the movie Star Wars. Robots in this galaxy, however, have many different functions and come in all shapes and sizes.

A **robot:** an automatically guided machine which is able to do tasks on its own, almost always due to electronically-programmed instructions. It has some ability to interact with physical objects and to do a specific task or a whole range of tasks or actions. It may also have some ability to perceive and absorb data on physical objects or its local physical environment, to process data, or to respond to various stimuli through **sensors**.

Photo: NASA

This is in contrast to a machine: a piece of equipment that uses physics to enhance the ability to do work. Simple mechanical devices such as a gear or a hydraulic press, which has no processing ability and which does tasks through purely mechanical processes and motion.

A computer is a special electronic machine that makes calculations and decisions. Many machines have computers on board that control the machine.

A robot is a super combo of a machine, a computer, sensors, and much more! While there are many different definitions of what exactly a robot is, many robots have a few common characteristics:

•Have the ability to sense its environment.
•Perform commands made by a computer program.
•Usually can perform actions, tasks, and/or movements.
•And much more!

DISCUSSION QUESTION: Is a dishwasher machine a robot? What do you think? Why?

____

____

Why Do We Have Robots?
How Are Robots Useful?

•There are many needs for robots in our society, each type carries out a different function to help humans take dominion over the earth and help mankind.

•Robots do great in jobs that are: **boring**, **precise**, & **dangerous**.

• There are many different types of robots, a few examples include **industrial robots**, **service robots**, **exploratory robots**, and **entertainment robots**.

•**Industrial robots** complete tasks that are very repetitive and precise, such as putting car doors on a car frame a few thousand times every day. Robots can perform a simple task far faster than a human can, leading to the production of quality products at a quick rate. They can also handle objects that are too fragile for a human to handle, and are now used in medical surgeries!

•**Service robots** can be used to do tasks such as robotic vacuum cleaners and lawn mowers, and to also retrieve and deliver objects, which could be helpful for elderly people.

•**Exploratory robots** and military robots do missions that are too dangerous for humans, such as going to Mars, to the bottom of the ocean, inside a volcano, and to search and rescue for trapped victims in an earthquake or landslide.

•**Entertainment robots** are produced for and by people interested in creating what is really a technical form of art and motion. It is a challenge to plan, build, program, de-bug, and test a robot to complete human-like tasks!

As the world is becoming more technological every day, the use of robots is increasing. More and more robots are doing things that can be used to help out humans. The field of robotics science is very exciting and constantly changing!!!

Mindstorms NXT™ Intelligent Brick

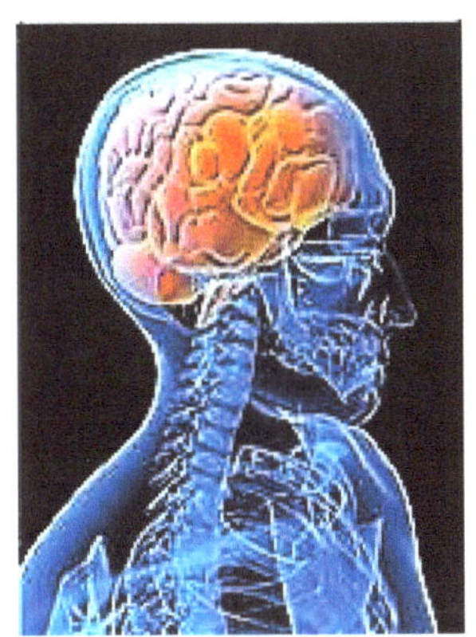

The Mindstorms™ set and pieces contains many complex parts. The main part is called the Intelligent Brick (IB). The IB is also called the "brain." It is a real LEGO® computer! (It is also VERY breakable and VERY expensive – so don't drop it!)

•You download programs from the computer to the IB via a 2.0 USB cable or via Bluetooth™.

Some features of the IB include:

•32-bit microprocessor.
•Support for Bluetooth™ wireless.
•1 USB 2.0 port.
•4 input ports.
•3 output ports.

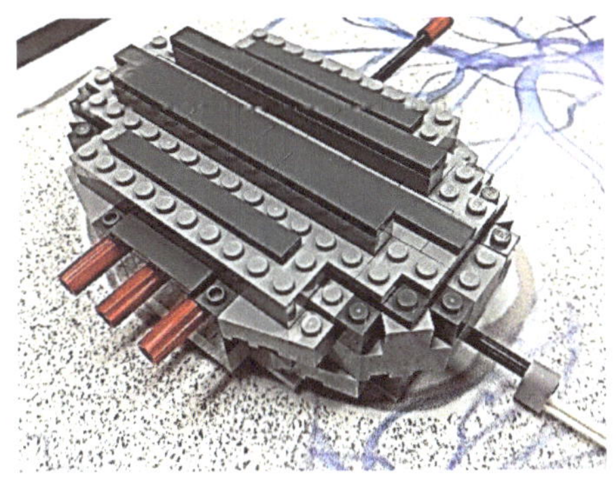

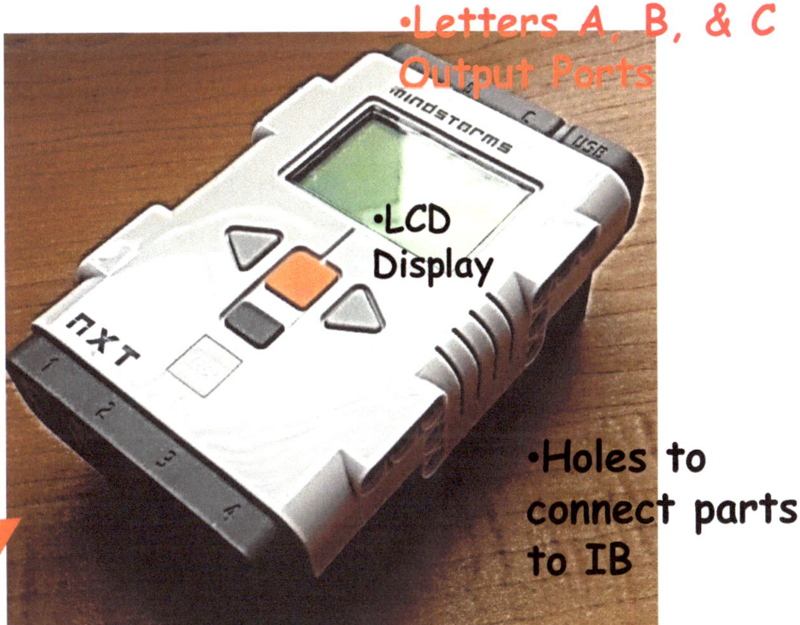

Intelligent Brick User Menu

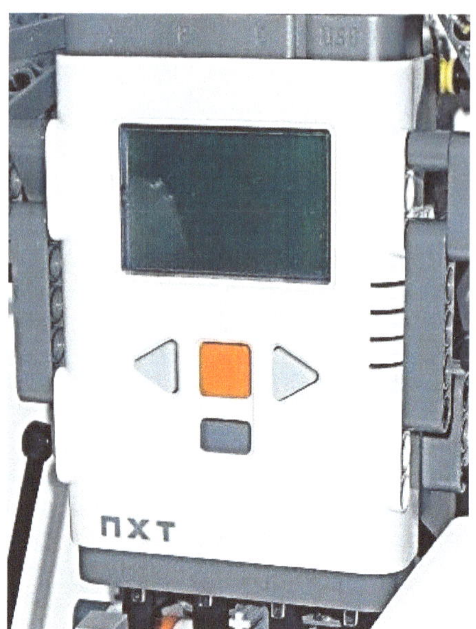

Button operations:

•The Large orange square is the "on" and "select" button.

• The two gray triangular shaped buttons are to navigate left or right.

•The rectangular dark gray button is "back."

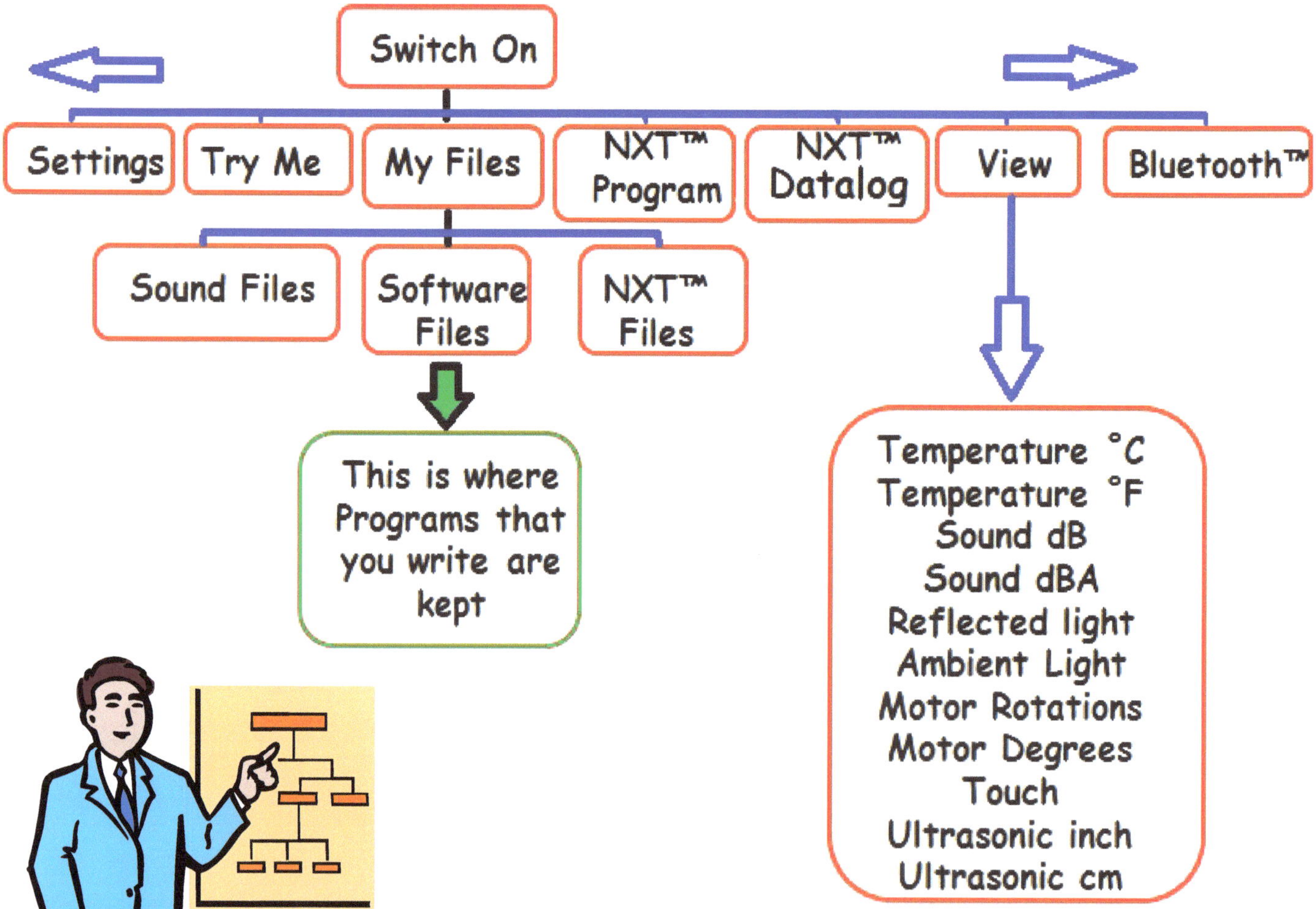

Switch On

- Settings
- Try Me
- My Files
  - Sound Files
  - Software Files — This is where Programs that you write are kept
  - NXT™ Files
- NXT™ Program
- NXT™ Datalog
- View
  - Temperature °C
  - Temperature °F
  - Sound dB
  - Sound dBA
  - Reflected light
  - Ambient Light
  - Motor Rotations
  - Motor Degrees
  - Touch
  - Ultrasonic inch
  - Ultrasonic cm
- Bluetooth™

Servo Motors

A robot also must be able to complete tasks or actions in its environment and usually has some ability to interact with physical objects. Mobile robots have the capability to move around in their environment and are not fixed to one physical location, and complete this motion and movement by the use of motors.

The motors that we will be using in this class on our NXT™ robots are called Servo Motors. They have an irregular shape, and have built-in Technic™ beams for easy connections. The servo motors are made very well, and can be programmed so that the robot's movement is very precise.

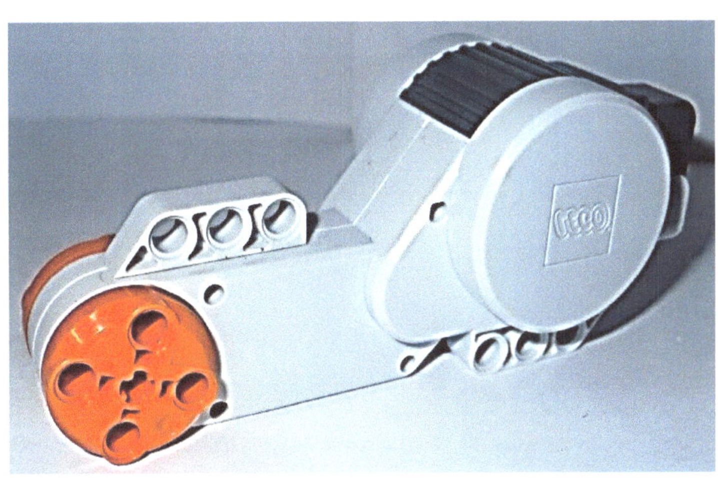

Servo motors also have a unique feature called the tachometer, which is a rotation sensor. It can measure in rotations or in degrees. You will see that this feature will come in handy soon!

Output Ports for Motors

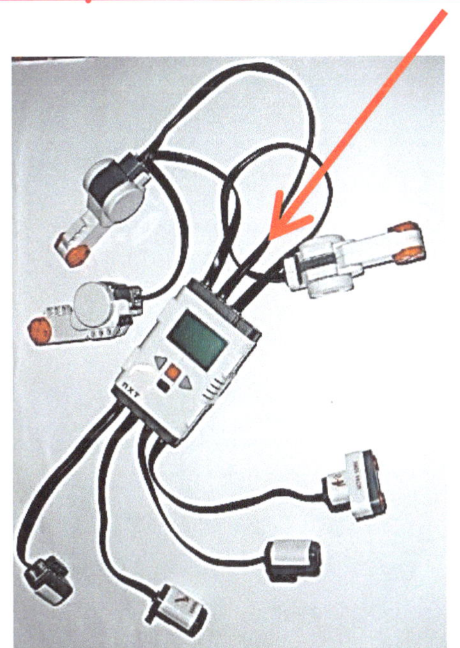

The name of the Output Port determines the name for the motor. For example, when you plug a motor & a cable into Output Port A, that motor is now called Motor A. This is important for programming!! If you switch the motor into Port B, that motor is now called Motor B.

LEGO® Robotic Pieces

Apart from the electronic elements that we will be using to create our robots, we will also have access to ~ 10,000 LEGO® pieces! Of course, all pieces NEED to be kept ORGANIZED and NEAT for efficient and precise robotic building!

We will discuss in class the different LEGO® Technic™ elements and parts. Each LEGO® piece has a specific name! Pieces are named according to the number of modules, or studs, that they have. Here is a chart for reference of just a few pieces that you will commonly use in this course:

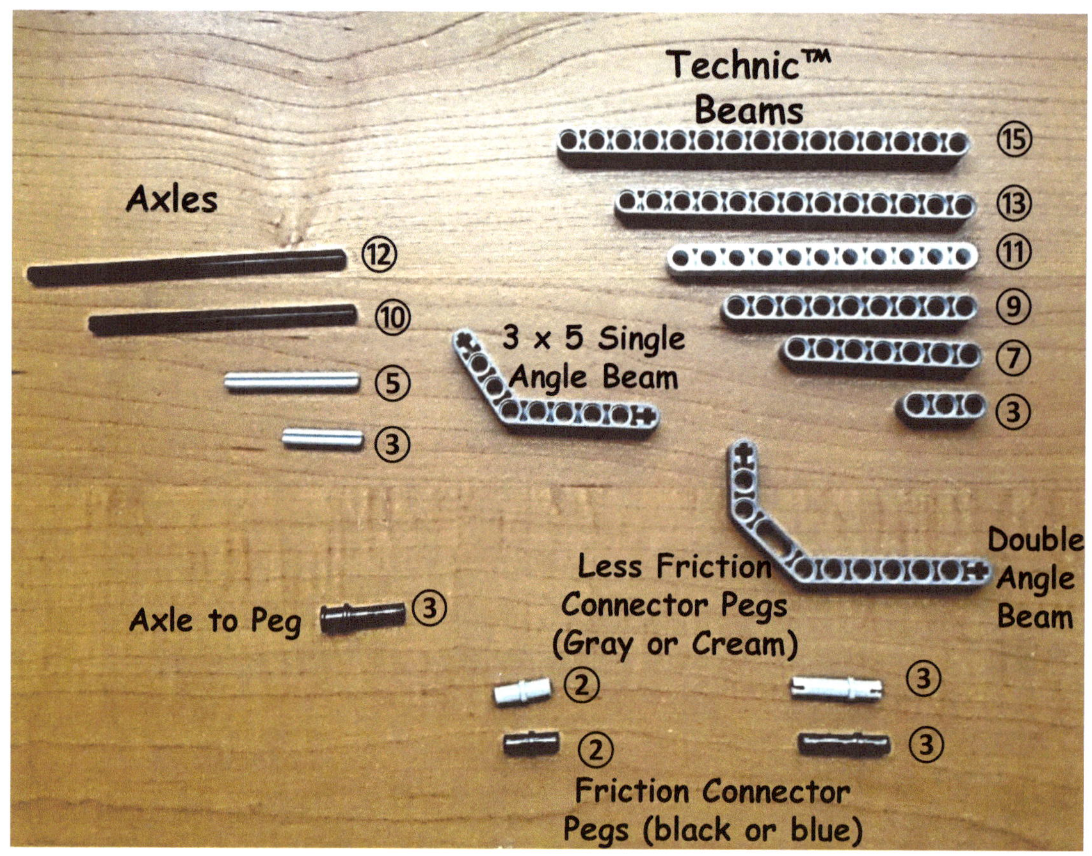

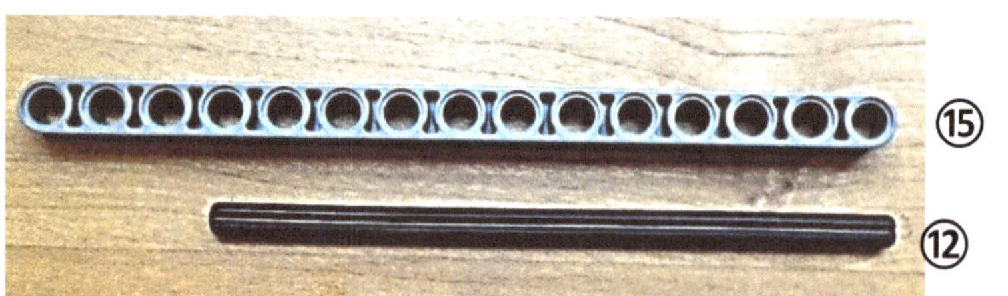

The sizes of the axles can be determined by matching it up to a beam. Count the number of holes to determine the length of the axle.

The Core Robot & Good Robot Building Features

•When solving missions, the most important thing is the <u>core robot</u> that you will use. Everything else that you do during class is added to the core robot, such as manipulators, programs, extra motors, etc.

When building a core robot, you have to keep a few things in mind:

1. The robot must fit in the area that you will do task missions.
2. The robot must be stable and well balanced, preferably with a low center of gravity.
3. Most of the robot's weight should be on the motorized wheels.
4. The best core robots are usually symmetrical.
5. The robot should be well built, strong, and solid - not falling apart.
6. It also needs many possible attachment points for manipulators.
7. It should be made with as few pieces and as simple as possible, to keep down on weight.
8. It should have a differential drive system.

•Our first robot in this class is named WheelBot, which is a very versatile robot that has all the above requirements. Please make sure that your tires are attached to the wheel hubs correctly.

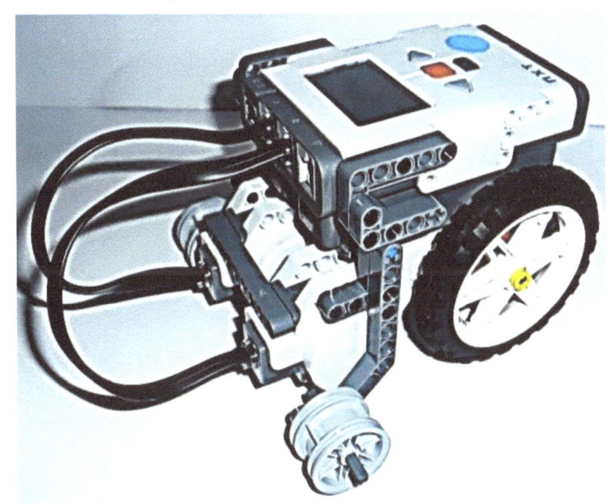

The Differential Drive System on Our Robots

It is important for robotics students to understand the differential drive system.

Here is an underside view of your robot

In this system, each working wheel is connected to its own servo motor, one on each side of the robot (B & C).

Each wheel works independently of the other.

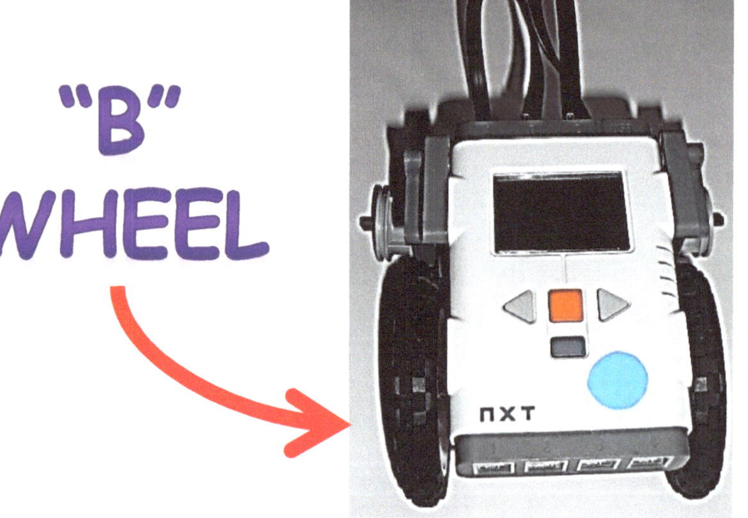

- If you program <u>both</u> motors to go, the robot will go either forwards or backwards.
- If you program only <u>one</u> motor to go, the robot will make a turn.
- If you program one wheel to go one direction, and the other the opposite direction, you will get a quick spin. You have to do this by adjusting the STEERING function.

GET THINKING!
Why are there two extra wheels that lack tires on our robot?

____

____

UNIT 2
INTRODUCTION TO PROGRAMMING

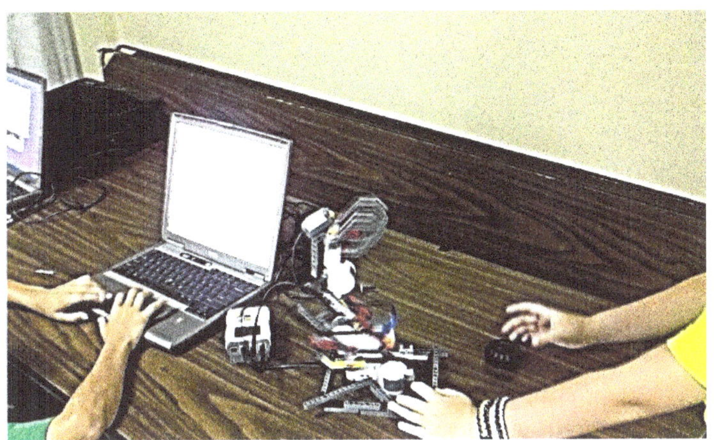

How do we get the robots to do what we want them to do? It all runs from the programming! The program, or commonly called the code, is what makes a bunch of hardware (in our case, LEGO®s) become ALIVE!

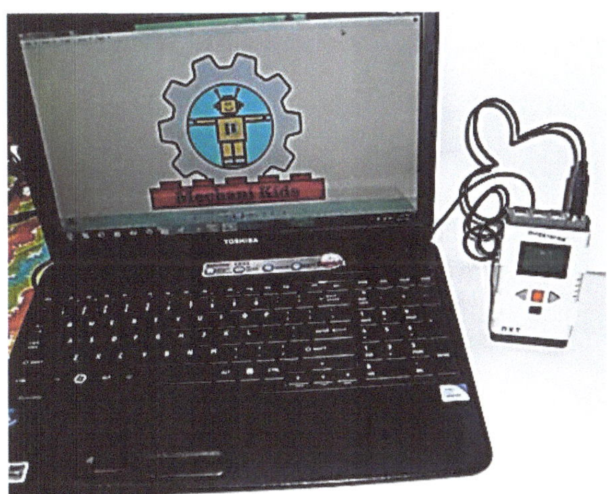

Intro to Programming

An important part of robotics is the computer programming. A **program** is **the set of instructions or commands that tells the robot what to do.** To program our robots, we use the robotics software on the computer to write the program in a specific **programming language**. The programming language we will use is called NXT - G, which is based on icon graphics.

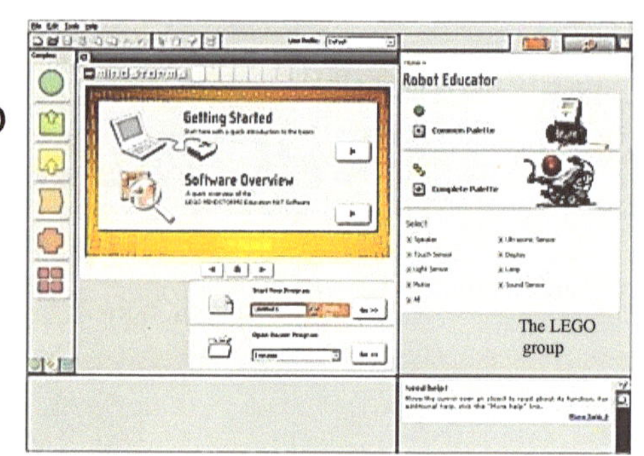

USB Plug

After you write the program, you download it onto the IB via USB wire, or by using Bluetooth™. Then, the "brain" of the robot **runs, or executes**, the program.

Learning to program is so important - the program is what makes a bunch of LEGO®s become ALIVE! Computer programming is a form of art and engineering. **Engineering is when you apply scientific and practical knowledge in a logical order to solve a problem by designing or improving devices or systems**. It is also a form of art, because you can be as creative as you want when making the commands to best execute the mission. Also, many programs that do a specific thing can be written in a variety of ways. It usually involves a lot of creativity and experimentation, a real work of art.

Sometimes programming can be frustrating and difficult, and many times the program doesn't run exactly the way you want it to. Usually this is because of an error in the program that needs to be corrected. If your program does not work correctly, look at it like a mystery or puzzle to find out what is wrong - and usually it is the programmer that makes a mistake when writing the program.

Qualities of a Good Program

When writing a program, you can usually find more than one way to complete a task mission. You will notice that as you program, you will develop your own unique method and style of programming - which is great! But all good programs should have the following 3 characteristics:

1) The program does the desired task.
2) It is easy to change or adjust.
3) It is understandable by you and others who knows the programming language - including me the teacher! If the program does the correct function, but you have no idea how it works, and I can't make sense of it, then it is not a good program! ☺

How Does a Program Really Work?

You first start off with what the description of the task mission. Then you break down the requirements, in a process called **flowcharting**. This is just figuring out all the steps and placing them in a certain order. During this process, you do the 3 C's (See page 21). Then you use the **programming environment** (NXT™ software) as a tool to create the program. The programming environment is also called the **IDE**: **integrated development environment**. You use a **programming language (source code)** to then write your **program**. Our NXT - G language is a graphical programming language, as opposed to a text-based programming language. Then the software translates the NXT-G or **compiles** it so that the computer can read it, and then it gets **downloaded** to the NXT™ IB. The program gets **executed**, and then you may have to **modify** and **debug** the program. **Debugging** the program is the process of finding and fixing errors.

Computer Software

Here are the major features of the Mindstorms™ software user interface:

1. Where you name/start a new program
2. Where you open a recent program
3. Tool bar
4. User profile logon
5. Little help window
6. Work area map
7. Programming palette
8. Configuration panel
9. Controller
10. Programming Workspace
11. Save file button

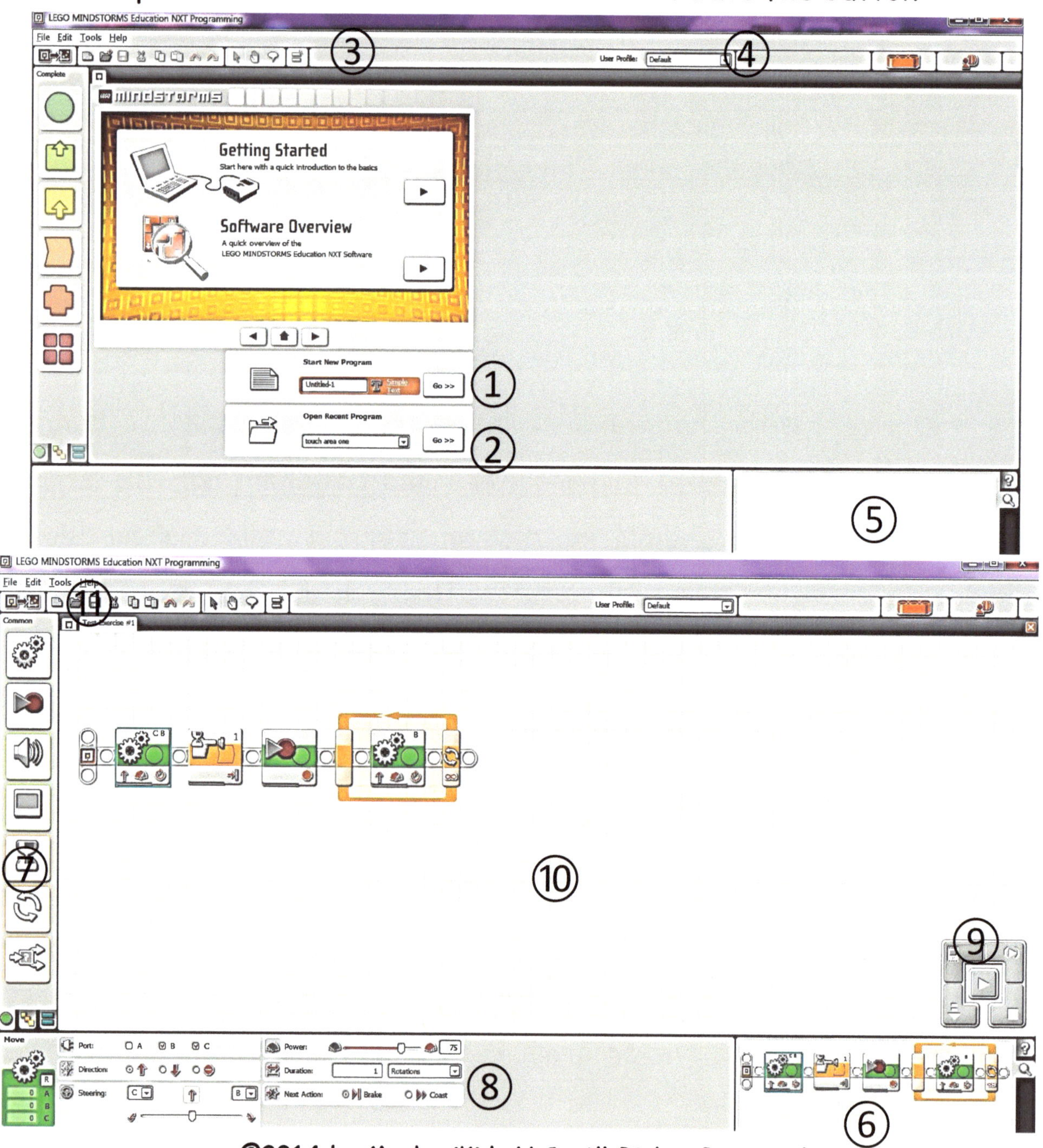

Making a New NXT-G Workspace

Before we start programming, you and your teammate need to make a user profile. This will just organize your programs that you create for the year and keep others from just opening up your work!

When you open up the NXT programming window, you should see this in the middle of the top user bar:

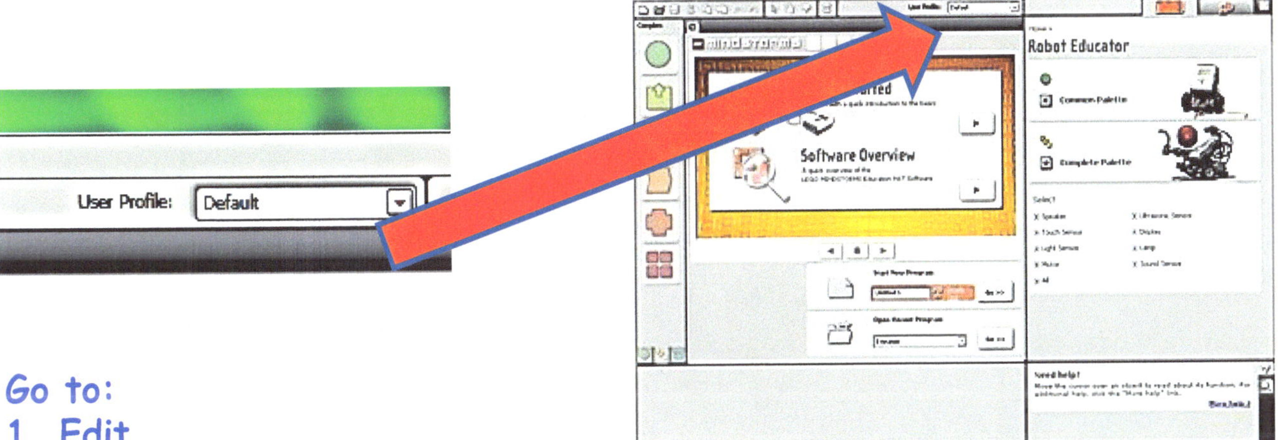

Go to:
1. Edit.
2. Manage profiles.

You should see something like this:

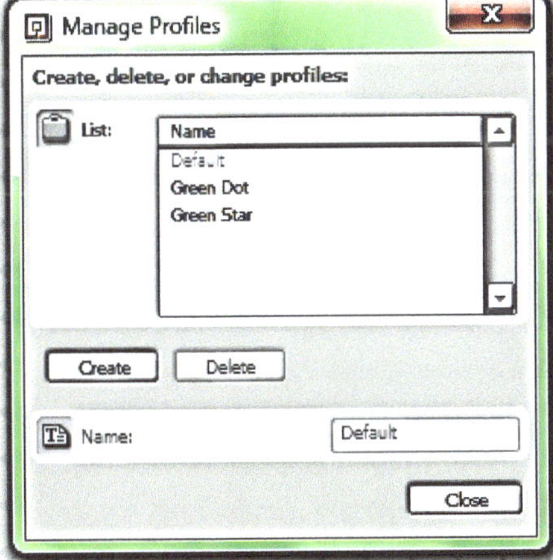

Go to "create", and then put in:
1. School year
2. Day of the week
3. Your class time (<u>use a period instead of a colon</u>)
4. Robot's name

EXAMPLE: "2014 Monday 1.30 Green Dot"

<u>*MAKE SURE TO ALWAYS SIGN IN TO YOUR CORRECT USER POFILE EACH CLASS TIME!!!!!*</u>

The Programming Palette

The programming palette contains all of the programming blocks you will need to create your programs. Each programming block includes instructions the NXT™ can interpret. You combine the blocks to create a program.

The programming palette you see under the tab of the green circle is called the **common palette**. We will discuss the other palettes later.

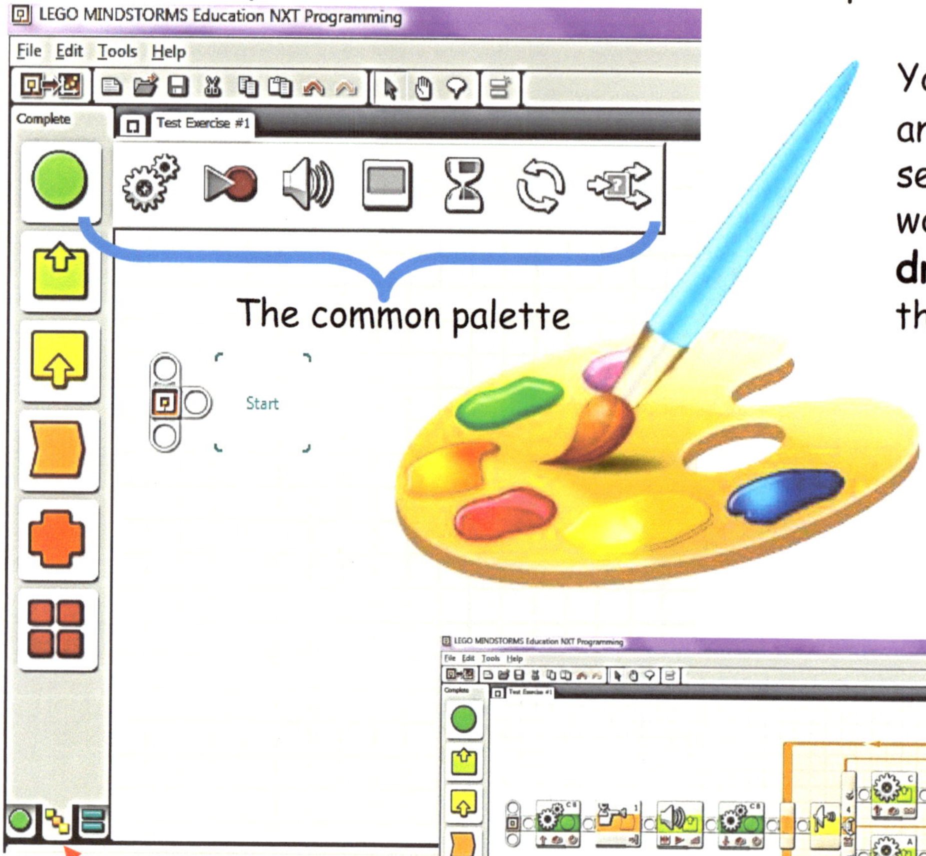

LEGO MINDSTORMS Education NXT Programming

File Edit Tools Help

Complete

Test Exercise #1

The common palette

Start

You press on a block and **drag** it to the sequence beam in the work area. Then you **drop** it into place in the program.

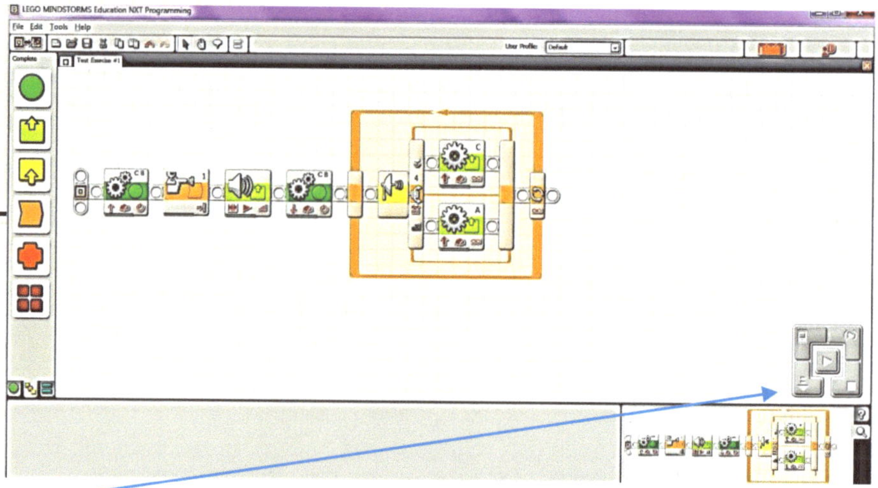

The complete palette. (The common palette is a part of the complete palette.)

Once the robot is plugged in with USB, and it is TURNED ON!, then you press download. YOU WAIT UNTIL YOU HEAR A BEEP FROM THE ROBOT AND THEN WAIT 10 SECONDS. Then you unplug the robot and take it to the mission field to run the test program.

Program Blocks of the Common Palette:

Move block
The Move block makes your robot's motors move!

Record/Play block
The Record/Play block lets you to program the robot by recording the movement of the motors turning, and then after recording, you can play play back the movement.

Sound block
The Sound block lets your robot make sounds.

Display block
The Display block allows you to control the LCD display screen on the IB. You can show icons or text data on the robot when running your program.

Wait block
The Wait block makes your robot wait for sensor input during the running of the program, such as waiting until the robot touches, hears, and / or sees something, or until a period of time has passed.

Loop block
The Loop block repeats all or a part of your program to make your robot do something again and again, such as moving forward and turning for 5 times or until a sensor sees your hand.

Switch block
Your robot will be able to make decisions with the Switch block, which really gives your robot autonomy.

Introductory Programming Exercises

Exercise #1: Forward 2 Rot.

Open up a new program, and now program the robot to go forward and stop. Drag a move block to the workspace and set it for B&C forward, 2 rotations. Then the robot should stop. Now press the download button. Unplug. Run the program on the NXT™.
Yay! You have written your first robotic program!
Write in the information for your programming block: Example:

Forward
Move
BC 2R.

Exercise #2: Go forward and backup

Make the robot go forward 4 rotations and back 4 rotations.

Exercise #3: Right hand turn

Make a right turn using only a single wheel.
Making the robot do a right hand or left hand turn is tricky! You need to take a moment to think before you program when you make any turns. Let's break it down. A circle can be broken up into 360 degrees. A right angle is 90 degrees. So if we want the robot to do a right quarter turn, we mean that it turns 90 degrees from its current location. This is making the robot turn 90 degrees in orientation - not wheel rotations.

We need to figure out how many degrees we turn the wheel to get the robot to move 90 degrees. Here is a little bit of circle/wheel information:

.To get the exact number of rotations of the wheel to get the robot to turn 90 degrees takes a little bit of math, which will be discussed in class.

TRY IT! Type in 90 degrees for the wheel to turn. Test run the robot. Does the robot make a perfect quarter turn? Then add or subtract degrees by estimating to find out how many degrees it takes to make a perfect quarter turn of the robot.

- One full circumference is one rotation of the circle/wheel = 360 degrees
- Half of a rotation is a straight angle of 180 degrees.

Exercise #4: Around the obstacle and back to base.

How many programming blocks will you need?________

OBSTACLE

Remember, when writing programs for your robot, come up with a plan that is as simple as possible. Your goal is accuracy and repeatability!

Commenting Code

•Sometimes when you make increasingly complex programs, you need to add additional information that explains what each part of the program language is doing. This is when you add descriptive (describing what it does) text to the workspace area that explains in English (or your native language) to let another person (or maybe even you!) understand what is going on in the program and flow. Comments also help you to know where you are in the program while creating it, so that way it is less confusing while writing out a complex program or when debugging it when testing the program.

You can leave comments in 2 ways:

1. An overall program explanation.
2. Making a comment for each program block

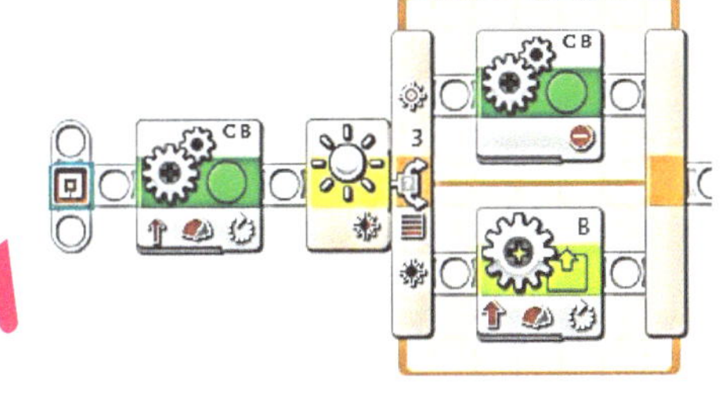

Overall program explanation:

You do this by pressing on the MINDSTORMS icon at the start of the sequence beam. There opens up a space on the configuration panel where you can type:

Info: This program start with a forward movement, and then checks the light sensor for light areas or dark areas. If it senses a light area, then it stops. If a dark area, then it makes a turn with B motor for 45 degrees at 75 power.

Comments for each program block:

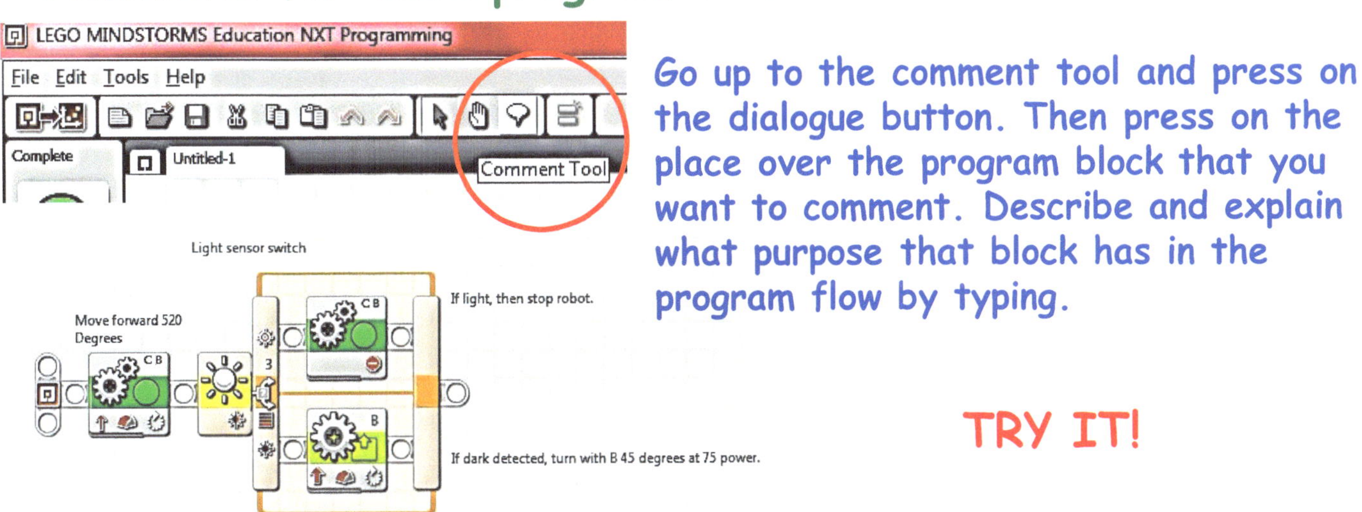

Go up to the comment tool and press on the dialogue button. Then press on the place over the program block that you want to comment. Describe and explain what purpose that block has in the program flow by typing.

TRY IT!

1. How to write an overall program explanation:

You do this by pressing on the MINDSTORMS icon at the start of the sequence beam. When you do this, it opens up a space on the configuration panel below where you can type a large amount of information in a paragraph:

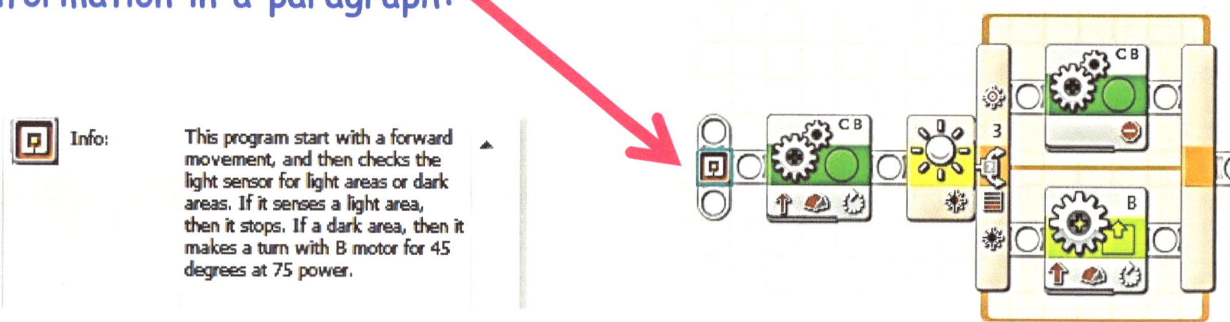

2. How to write comments for each program block:

Go up to the comment tool (see red circle) and press on the dialogue button. Then press on the place over the program block that you want to comment on. Describe and explain what purpose that block has in the program flow by typing directly on the workspace.

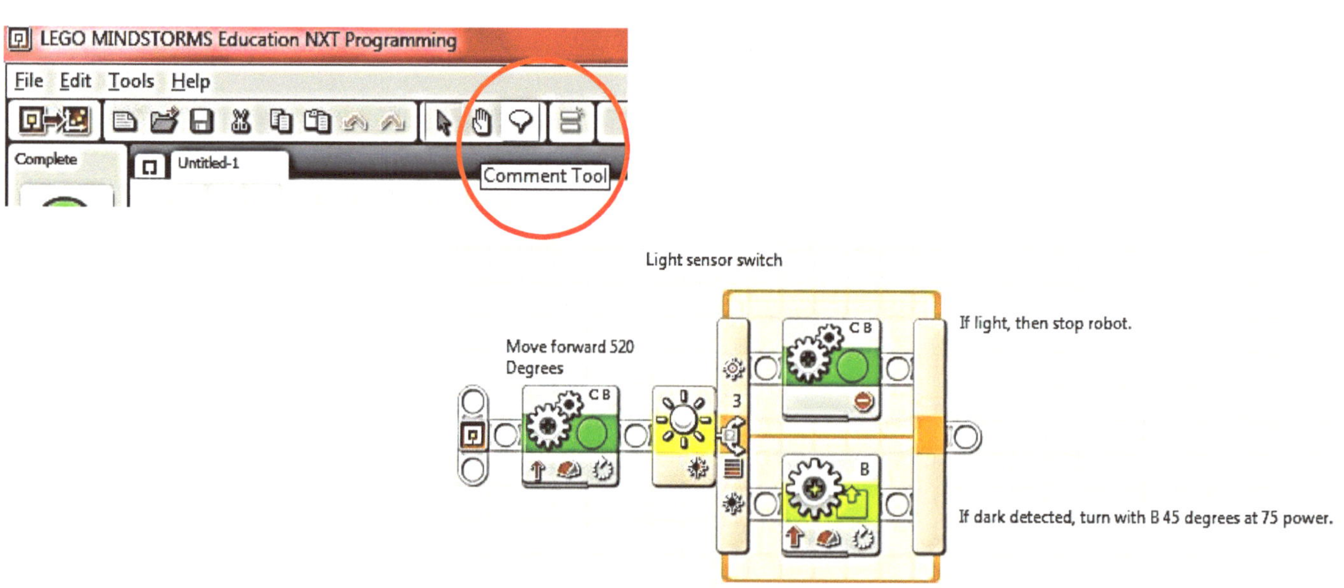

UNIT 3
ROBOTIC MISSION PLANNING

When completing robotics assignments, like many things in life, the end of a great mission or achievement starts with a good, solid plan.

Flowcharting

Flowcharting is a process that has the person placing steps of a process in a logical order. Just like when we build LEGO® robot builds step-by-step, you make your plans with each step in a sequential order. Below is a diagram of a flowchart where you can visually represent a plan:

Linear Flowcharting Exercise:

Intro to Tasks

Task Missions are robotic missions that your team must work on together to solve.
A bunch of task missions make up a Challenge.

A task mission is solved by doing the 3 C's:

The 3 C's:
Course, Code, & Components

Course- this is the actual route that your robot goes to reach the target areas to complete the mission. Always draw out EACH movement that it will take to get to the mission target and back. It helps to have a drawn out map of the mission field to help you figure out your course.

The best courses are simple, easy, and with as many straight lines as possible. You want the path to be the fastest for the robot to travel.

Code- the computer program you write with commands. You use your course and **flowchart** the steps of the mission, and then write the code using the NXT - G programming language to get the robot to carry out its tasks along the planned course. The best programs are made as simple as possible.

Components- These are additional LEGO® pieces you add to the core robot to carry out its task. When you add these additional pieces, it is called a **manipulator**. The manipulator must be made to complete the task, nothing else. Again, the manipulator needs to be made simple, and with as few pieces as possible.

8 Steps to Solve a Robotic Task Mission

When solving the task missions, there is a certain procedure a robotic engineer must follow:

1. You must get the directions for the task mission and then figure out the objective (goal) of that mission. Answer the question, "What do I need the robot to do in this mission?"

2. Come up with ideas, and formulate a plan for the mission with your teammate:

   BY:
   A. Think about the course, the code, and the manipulator (components) you must make to complete the mission.

   B. DRAW your planned course on a paper map!

   C. Make a FLOWCHART of the mission from the drawn course path.

3. Now you actually take the measurements for the path, build manipulators.

4. Now you write the program code on the computer.

5. Test the mission.

6. Observe problems and change course, code, and components as needed.

7. Re-test your mission.

8. To be considered a successful run, it must be valid and reliable. It must complete the run with success 2 out of 3 runs.

The Major 3 Types of Missions

As you progress through this robotics engineering course, you will face many types of missions. They can be quite complicated, and require a carefully made course, program code, and a specific manipulator component. Even though the task missions can get quite complicated, almost all task missions can be classified as one of the 3 major types of missions:

There are 3 major types of missions:

1. Strike
2. Delivery
3. Gathering

A Strike Mission is when the robot goes out in a mission field and bumps/strikes a target.

A Delivery Mission is when you take an object from base to a target and leave it there.

A Gathering Mission is when the robot goes out to the mission field and brings an object back to base. An object can be any size or shape, depending on what the mission is.

- Most of all the other task missions are a combination of two or all three of the above. You usually need to build a manipulator to do the missions, and you attach this manipulator to the base robot.

How To Solve Missions and Mission Problem Troubleshooting

As you may have noticed, problems are constantly arising during the mission completing process. Just when you think you have the entire mission fixed, a new little issue arises! This is true engineering, and just a part of robotics mission solving. There are, however, ways to make it easier to solve these problems.

The best way to minimize problems, however, is to start off with a good solid plan. When solving missions, ask yourself questions when planning a course, writing a program code, and building the components. Here are helpful questions to ask yourself during the process:

Checklist for Robotic Troubleshooting

1) Did you start off with a good flowchart first?

2) Check Hardware:
- ✓ Is the base robot the best for the mission?
- ✓ Are there component problems? (wheels attached properly, LEGO® pieces attached?)
- ✓ Are all the cables to motors and/or sensors attached properly and to correct ports?
- ✓ Manipulator attached properly?

3) Course:
- ✓ Is the course as simple as possible?
- ✓ Do you use lots of straight lines?
- ✓ What edges, obstacles, or walls can you use as a guide to realign the robot?
- ✓ Draw course path lines from the base to targets on your MAPS!!!!

4) Code / Program:
- ✓ Did you change the motor or sensor port to the one you need on the program block?
- ✓ Is the configuration panel adjusted correctly for each programming block?
- ✓ Are the programming blocks locked on the sequence beam? (not grayed out)
- ✓ Can a sensor help to eliminate making tons of exact tiny movements?

5) Sensor Use:
- ✓ Am I using the best sensor for this task mission?
- ✓ Did I make careful measurement sensor readings?

Common Problems

One of the most common problems with having your robot complete mission exercises is making correct and accurate turns. Using motor blocks really helps for making turns.

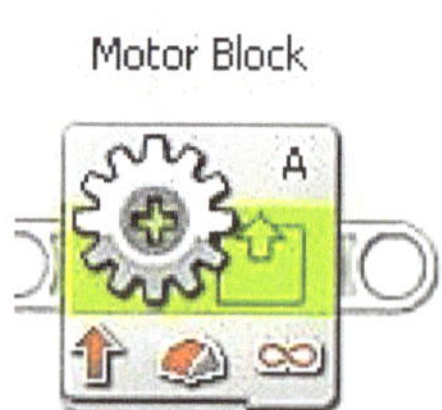

Another huge problem is making the robot go in a straight line. This can be caused by a variety of reasons, such as a badly-built robot. After checking to make sure your robot is stable and well-put together, there are different ways to find out exactly what is going wrong with your robot's movements. Using Move blocks are the best way to program your robot to make a straight line.

Move Block

•Exercise: Write a simple program: Straight Line Test.rbt. Program your robot to go all the way down the mission table. Measure the starting and stopping points from the side of the table. Does it make a straight line?

Most likely, your robot will veer off. There could be many reasons for this. Let's look at a few causes:

- •Wheels/ Tires not properly attached
- •Robot build: Skid Bots and Caster Bots
- •Type of skids or castor wheels
- •Servo motors

•Wheels: Most likely, your wheels are the problem. Take your wheels off your robot's axles and build the sample model in the picture. Then roll it down across the table. Does it roll straight or veer off to one side? You may need to change the wheels.

Hardware Solution Ideas

•Robot Build: Skid Bots / Caster Bots: The robot itself could be the problem. If you use a caster bot, the wheel could jog the position at the start of the run. Also: is there enough weight on the wheels so that they make contact with the floor and don't skip?

•Type of skids or caster wheels: Some skids work better than others. Try the different ones below to see if they skid along well both forward and back and perpendicular, such as when making turns. Try the different ones in the picture to the right. Also: what if you build a model that uses a ball for the caster wheel?

•Servo Motors: The servo motors could be turning at different rates. The Move blocks are supposed to help correct this, but sometimes it is off a little bit. You might need to find motors that turn at the same rate.

UNIT 4
ROBOT MOVEMENT & MANEUVERS

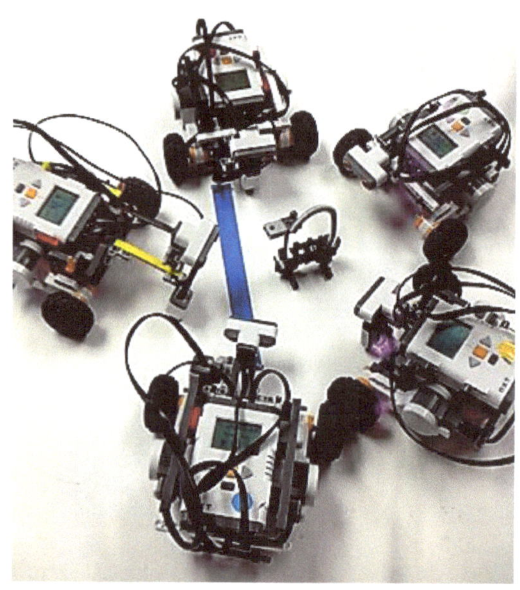

In this unit, you will learn how to get the robot to make movements precisely and efficiently. Hang on, because this unit is all about robot locomotion!

Photo: NASA

Wheel Duration

♫The Wheels on the Robot Go 'Round and 'Round! ♪

•The pros and cons of using rotations/degrees/seconds:
You will notice that when you go to the configuration panel of a move block when programming, you will have a choice of setting the motors using **rotations, degrees, or seconds**.
What's the difference?

•Rotations are useful when going long distances, but can be difficult to use for precise, small movements.

Full Rotation

•Degrees are very precise, but are hard to calculate over long distances.

•Remember that 1 rotation = 360°

•Seconds are useful if you want the robot to go to a location that is against the wall or to bump off an obstacle.

Why? If you program your robot to go, say, up against the far wall and then come back, and you mistake the distance as closer than it really is, your robot will get stuck because it can't finish the rest of the program until it completes its programmed rotations. Your robot will keep trying to spin its wheels until it completes its rotations before going on to the rest of the following programming blocks. If you use the seconds, it will let the time go by, and then keep going on to the rest of the program. You can also combine rotations and seconds by using two programming blocks.

Basic Robotic Maneuvers

Exercise #5: Timed Stop.rbt

Program your robot to move forward for 3 seconds and then stop. Measure the distance using a measuring tape from the starting point to the front of the robot.

Write Code Flowchart: Rate: ________ inches/4 seconds

Exercise #6: Gentle Bump Target.rbt

Have your robot move forward and bump into the target (gently) and then come back to base.

Target

Write Code Flowchart:

Please make sure that you understand the difference between degree turns of the wheel vs. degree turns of the robot's orientation.

If you want the robot itself to make a turn around a corner, say move to a position at a perpendicular angle, you do not just plug in 90° !!!

If you plug in 90°, what are you actually doing?____________________

How many wheel degrees makes the <u>robot</u> turn 90°?____________________

Robotic Task Mission Exercises

Exercise #7: Push LEGO®s.rbt

Push LEGO® blocks off target area in the mission field.

Drawing of Course: Drawing of Manipulator:

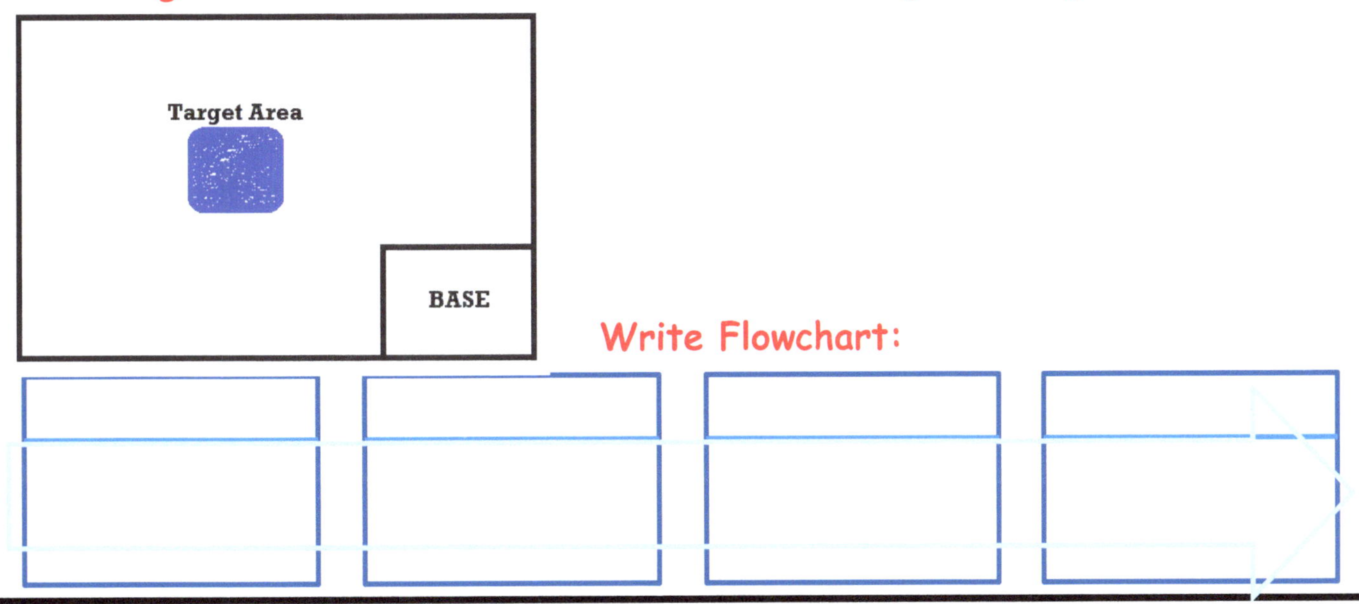

Write Flowchart:

Exercise #8: Send robot to go out and touch Area 1. Striped box is an obstacle.

Drawing of Course: Touch Area 1.rbt Drawing of Manipulator:

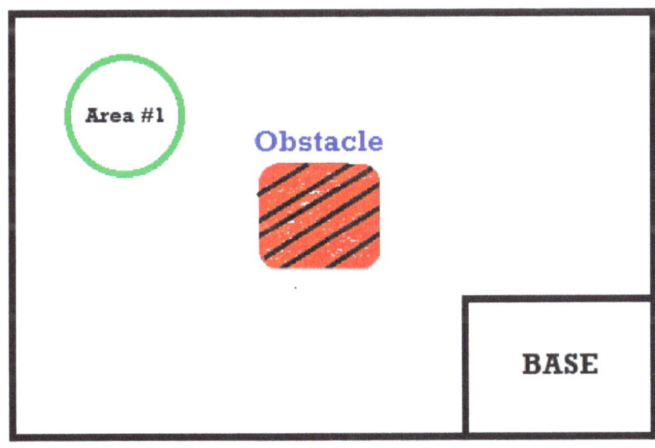

Write Flowchart:

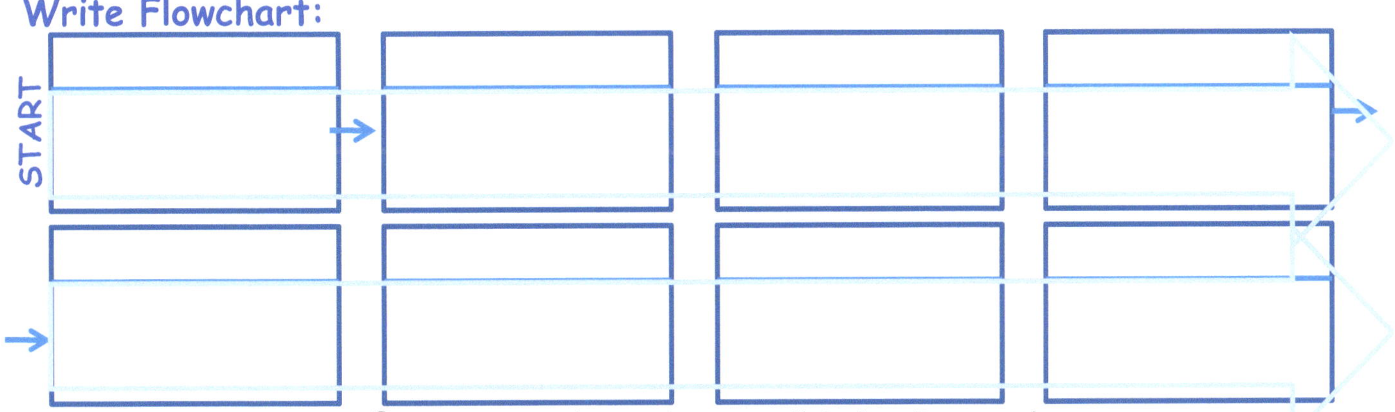

Making Measurements with the IB

When solving missions, it is important to make your robot travel distances to target in exact rotations / degrees / or seconds. When going far distances, remember, rotations are usually the best. When going short distances or making turns, degrees are the best.

Instead of just randomly guessing at the rotations or degrees on the mission field, you can actually measure it with your robot using the View tool. This is so handy and useful!! Here is how you get there on your NXT™:

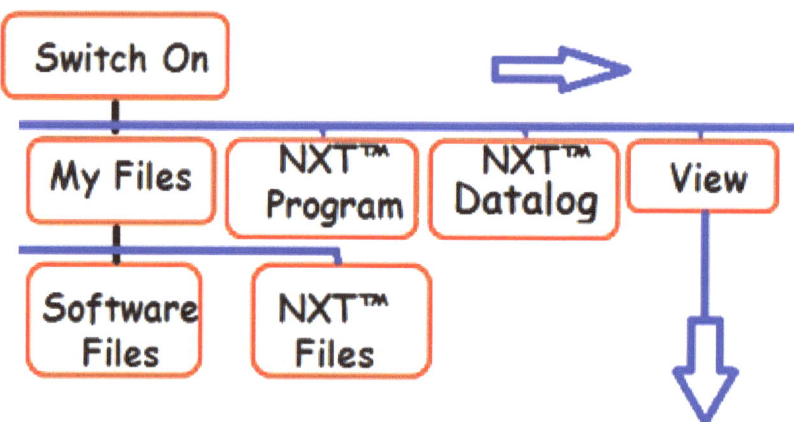

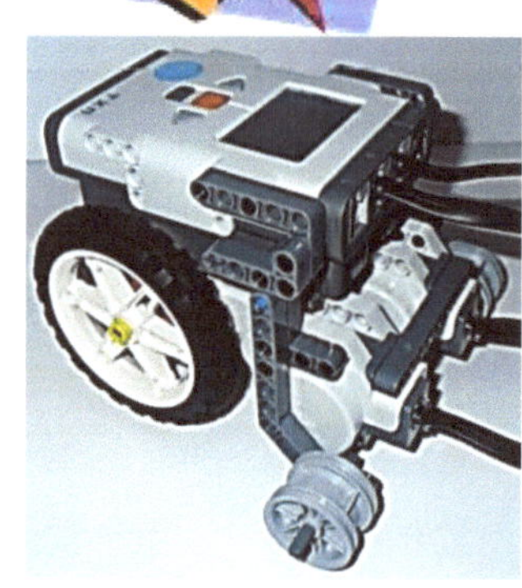

Temperature °C
Temperature °F
Sound dB
Sound dBA
Reflected light
Ambient Light
Motor Rotations
Motor Degrees
Touch
Ultrasonic inch
Ultrasonic cm

You pick which motor you want to measure (A, B,or C) and place your NXT on the starting point. Then roll/drag gently the wheels along on the mission table until where you are going (You may have to push the wheel along with your fingers).The LCD display should show the number of degrees or rotations that you traveled. If you are measuring a straight distance, which wheel should you measure? ______________________

APPLIED MEASURING: Roll the robot backwards while taking a rotations measurement in the View function. What happens? How do you program this? ____

Exercise # 9: Brick Retrieval Mission

Make your robot go out in the mission field to <u>retrieve</u> two bricks out of Area #1. The bricks need to be completely across the plane back at base to be considered successful. The robot is not allowed to touch the obstacle. The goal is to retrieve both bricks and have the robot back at base within 30 seconds. SAVE your program!

Drawing of Path:

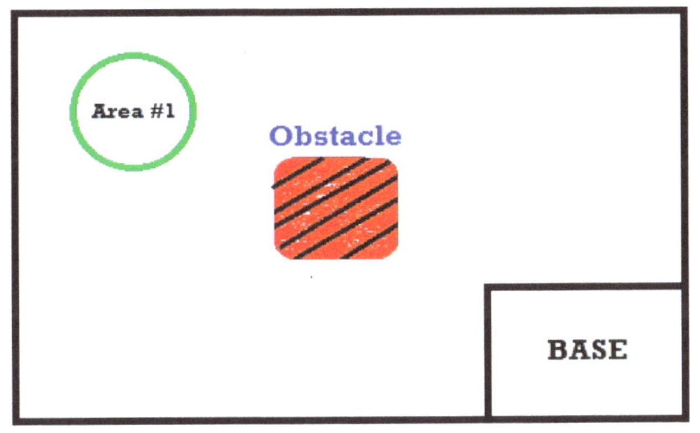

Drawing of Manipulator:

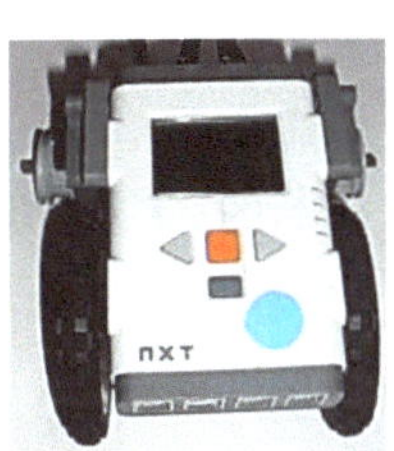

Write Program:

Exercise #10: Make your robot go out in the mission field to <u>deliver</u> one brick out at Area #2. The brick needs to be completely across the plane of the edge of Area #2 to be considered a successful delivery. SAVE your program on the computer!

Drawing of Path:

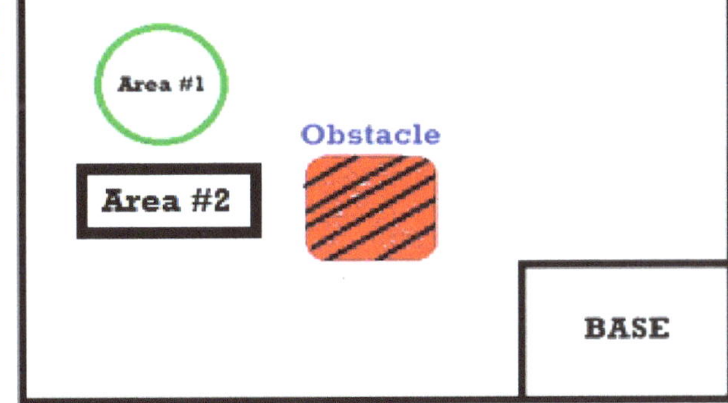

Drawing of Manipulator:

Robotic Turn Movements

•There are 3 different ways to make the robot turn:

- •Pivot Turn
- •Curve Turn
- •Swivel (Spin) Turn

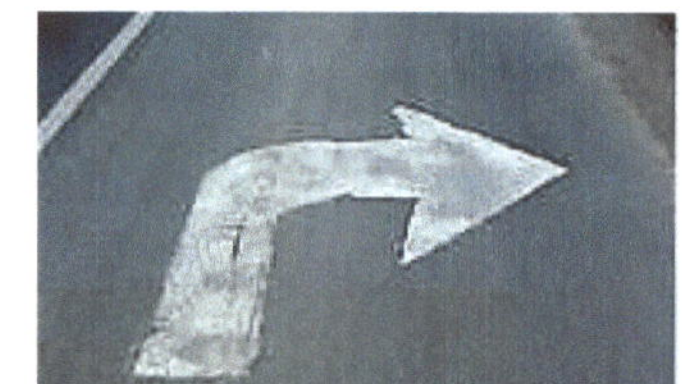

Pivot Turns

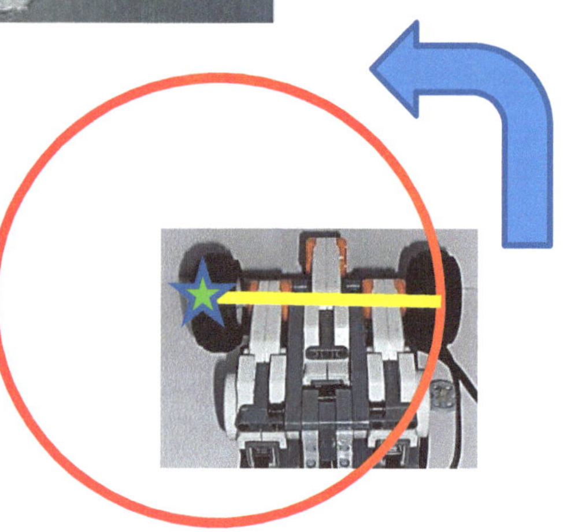

The Pivot Turn, also known as a single wheel turn, uses only 1 wheel. This type of turn is sharp and precise. The other wheel remains stationary. This will create a pivot point on the stationary wheel, with the track as the radius of the robot's chassis.
What type of program block(s) can you use to program a pivot turn?

____

Curve Turns

A curve turn is a gentle turn, where both motors are moving, just one is going faster and farther than the other. To make a curve turn, you adjust the steering on the configuration panel of a Move block. These are usually completed by trial and error to figure out exactly how far to move the slider bar. Make sure that you use the correct drop-down motor choices. What type of program block(s) can you use to program a curve turn?

____

Swivel/Spin Turns

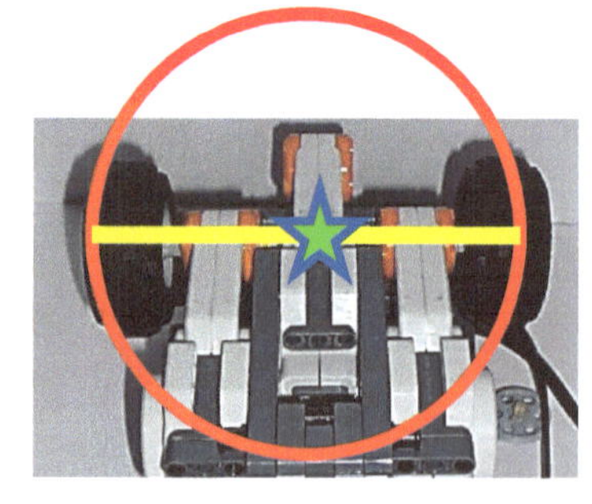

The Swivel/Spin Turn, uses both wheels of the differential going in different directions. This allows the robot to turn in it's own space and takes less room. What type of program block(s) can you use to program a swivel/spin turn?

____

Move Blocks

When programming robotic movements, you always use Move program blocks when having a 2 wheeled robot go forward and back. You can also turn off a motor on a Move block to make the robot make a pivot turn. The Move block is great for movements because it has the built in specialized capability to synchronize the two motors. The Move block has 2 GEARS on it!!!

Move Block

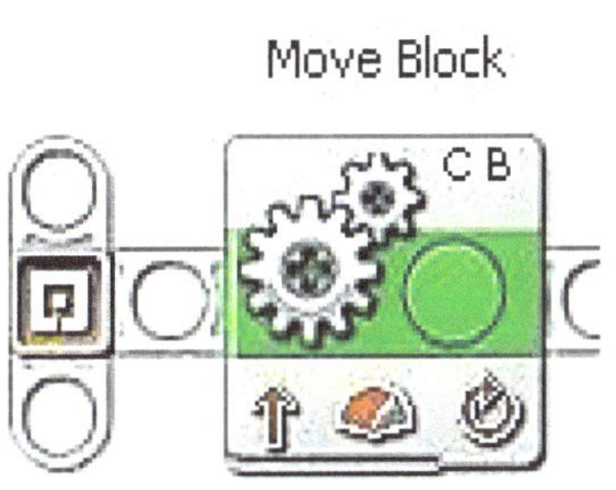

Here is the configuration panel for the Move block:

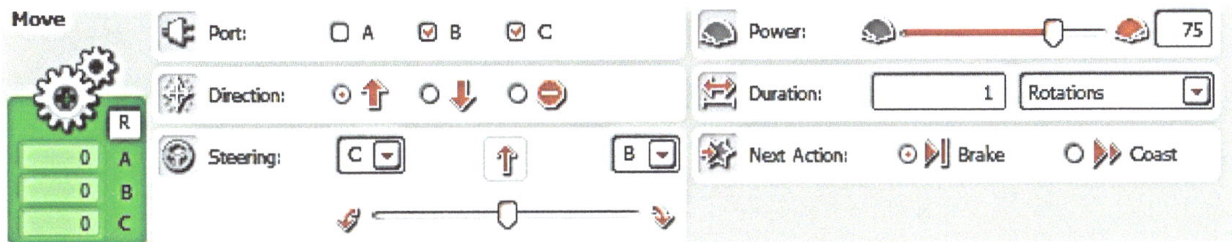

Move Block Steering

You can also adjust the steering on the Move block to make a curved movement or curved turn by adjusting the slider bar. These are usually completed by trial and error to figure out exactly how far to move the slider bar. Make sure that you use the correct drop-down motor choices. You can also have the robot spin or swivel, which is when one wheel is going forwards and the other backwards simultaneously.

Move Block Steering Exercises:

1: Have your robot make a gentle curved turn to the left. DRAW the path the robot took below:

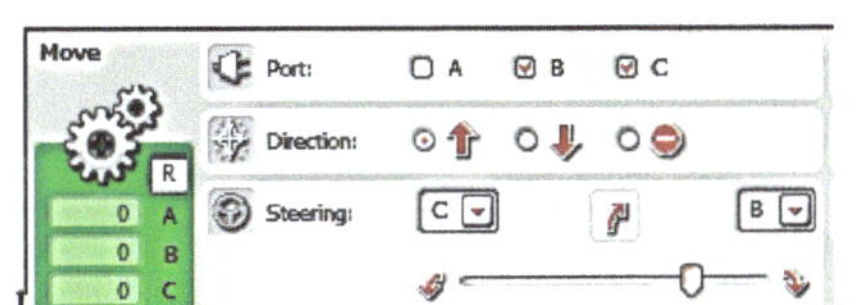

2: Have your robot make a sharp spin or swivel turn to the left. DRAW the path the robot took below:

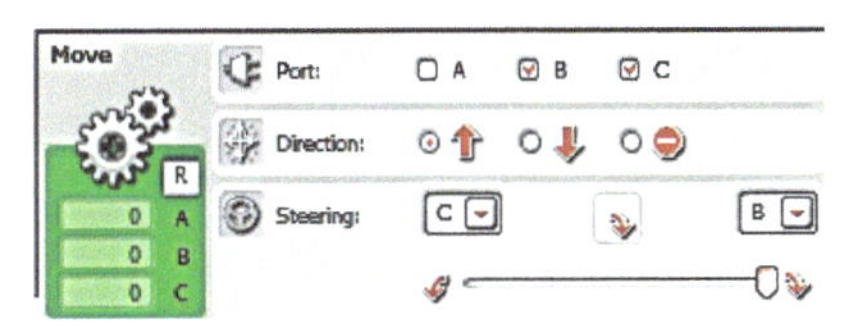

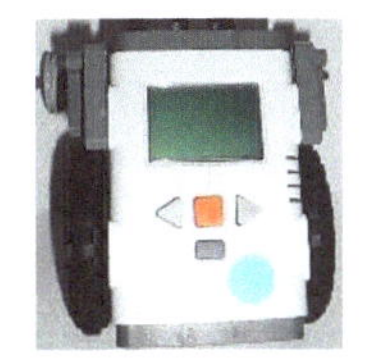

Motor Blocks

The Motor block is an action block used for controlling motors. It has 1 GEAR on it and it is a light green color. It is not in the common palette, but a part of the complete palette. This block only controls one motor at a time, and thus is great for very specific, precise, and finely-tuned motion action of one motor. Try the exercises below to learn how to use and adjust this programming block when making turns.

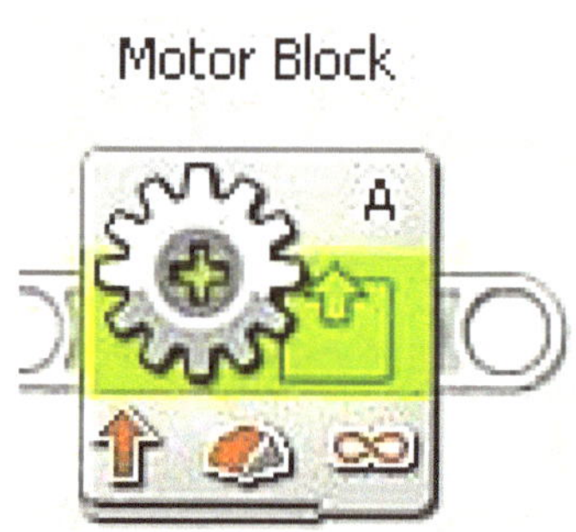

Here is the configuration panel for the Motor block:

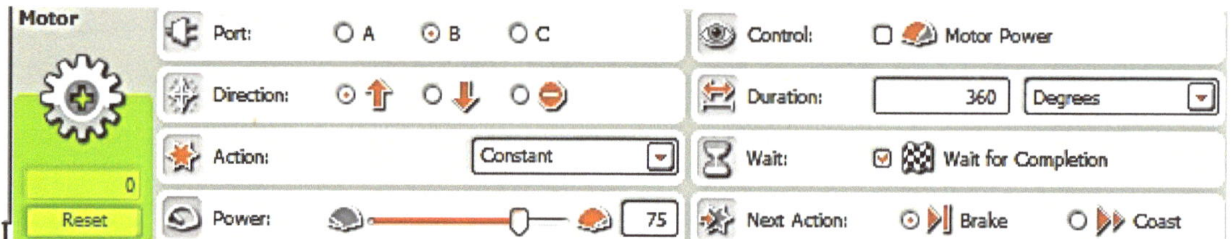

Motor Block Action Adjustments

The best part of the Motor block is the Action setting parameter. There are 3 choices:

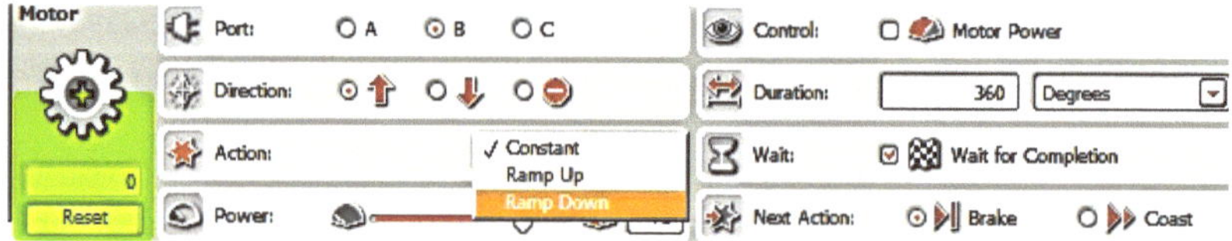

Motor Turn Exercise: Have your robot make a quarter turn to the right. Observe and record what each action setting did:

Constant:______________________________________

Ramp Down:____________________________________

Ramp Up:______________________________________

Record / Play Block

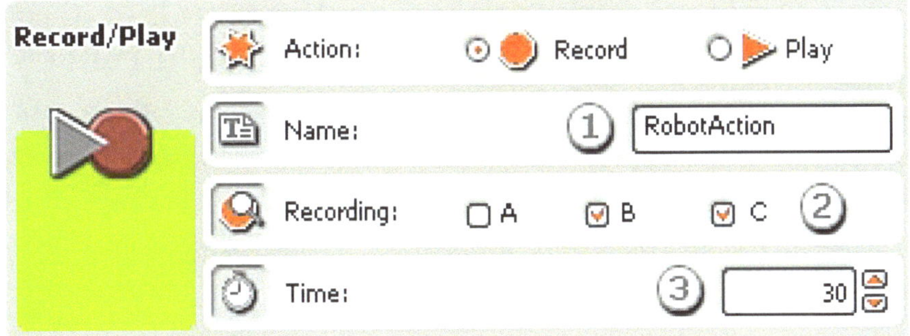

This programming block records an action movement that your robot does by calculating the movement of motors as you act out a robot maneuver. After recording the movement, it then can remember and re-play back that movement.

WARNING! It is NOT very accurate at all, but is a fun experiment to try and can be used for basic and uncomplicated movements.

Recording Actions:
1) Come up with a name for the action
2) Estimate in time how long it will take to complete
Example: You can call the file "Around Cone" and set the time to 8 seconds.
3) Now download the block. Run the program on the robot and act out the motion you want recorded. It records the motion that you acted for the time period that you had programmed.
4) Now, plug back in robot. Change the programming block's action settings from "record" to "play," and type the name you made up for the recording.
5) Download the record/play block again and run the program.

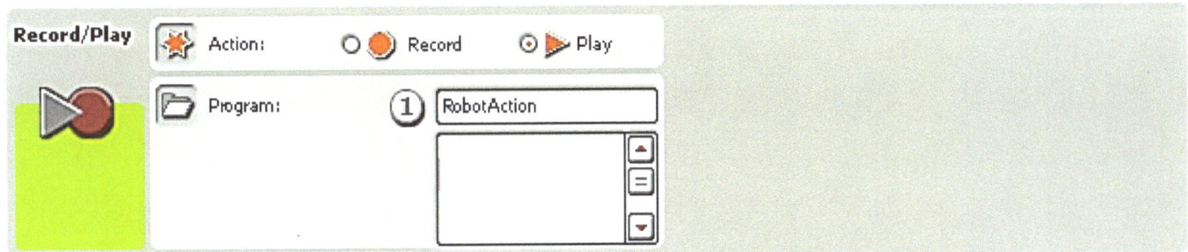

Sound Blocks

The sound blocks play a tone or a sound file OUT of the speaker on the Intelligent Brick. Sound and noises are what give a robot character, for example, imagine what R2D2 would have been like without all of the cute chirpy noises and beeps! Look at the symbols on the sound block. What does each stand for?

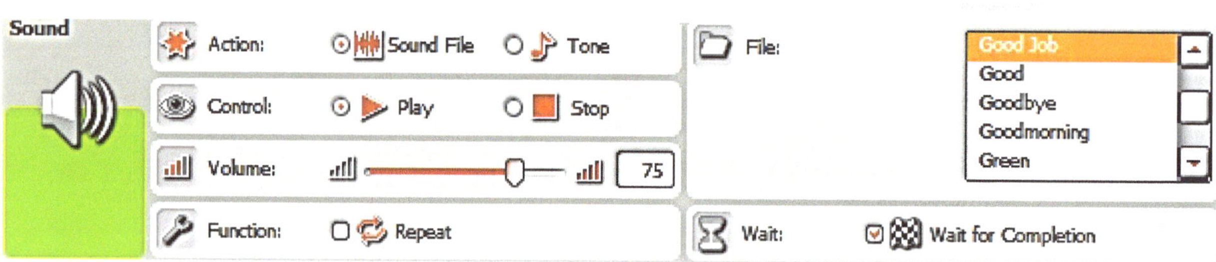

If you change the action setting to "tone," the configuration panel will look like this (below). By placing a bunch of sound blocks in a row on the program (with each set to play different tones), you can play a melody!

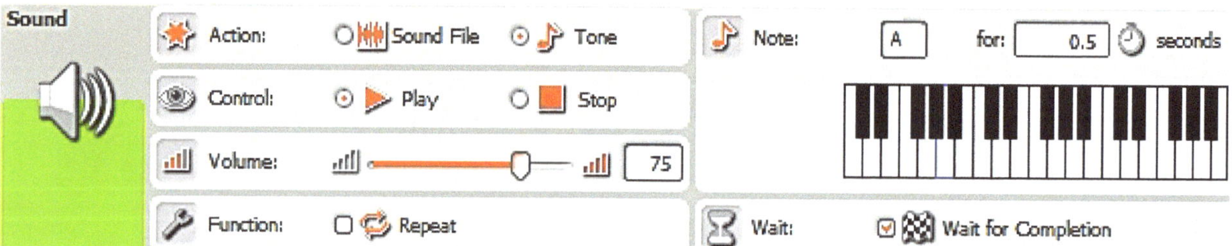

I'd like to draw your attention to the Function parameter: you can set sound files for the "Wait for completion" in the configuration panel. This will cause the sound file or tone to finish playing before the program moves on to the next program block. If you don't check this, the sound file or tone will continue to play while is goes ahead and moves on to the next block of your program. You can also play the file continuously by pressing "Repeat."

Sound Exercise: "The Music Sound Machine.rbt" Have the robot play a sound FILE and <u>then</u> move forward 1 rotation. Then have the robot make a sound that <u>repeats</u> while it backs up and then stops moving and also <u>stops</u> playing the sound. Then, make a quarter turn, and play a melody of 3 tones.

Turn Mission

In this mission, you are going to trace out a figure "8" around markers. You must use all 4 turning methods to turn around the markers - and don't touch them! DRAW the course that your robot took around the markers below. I would recommend you slow your speed during this exercise!

Pivot Turn:

Swivel/Spin Turn:

Curve Turn:

Record / Play Turn:

Robotic Self-Realignment

When you start your mission run, you must make sure that you start off from the same starting position each time so that your robot hits the targets. You will notice, however, if you repeatedly run the same mission, your robot will usually end up slightly off in a different ending place. This is because your robot may make a slip during a turn, hit a bump on the floor, or a knot in the wall when wall-following. The best way to overcome this is to have a point in the mission when the robot uses its environment to realign itself.

Self-alignment spots could be a line on the mission map, or an item on the mission field that is stationary, or it could be the table itself.

Squaring Up

Using the table walls is a great way to realign the robot in the middle of the mission run. We have used them in wall-following, but now we can use them to move away from the wall in a perpendicular motion. You must have a robot that has a side or edge that is flat, like this robot here. The side of the robot must not have anything protruding out past the edge of the chassis. There must be the ability to make a flush contact with the wall.

This type of squaring up is called *passive squaring up*, because there are no sensors involved and the robot just simply makes physical contact by bumping into the wall. The goal is to make a nice, soft touch with the wall and then get on its way.

Squaring Up Exercise

Write a program that practices using the passive squaring up method. Call this program "Wall SQ up.rbt." Have the robot go 6 rotations, turn 90 degrees in orientation, and back into the wall for 2 seconds. Then have it go forward 1.5 rotations and then stop. Then, try it with the robot backing into the wall at a small angle (turn not equal to 90 degrees). Did it work?

DRAW the diagram of your robot's path:

(robot table walls)

Maneuver Exercise: Red Ball Turn Run

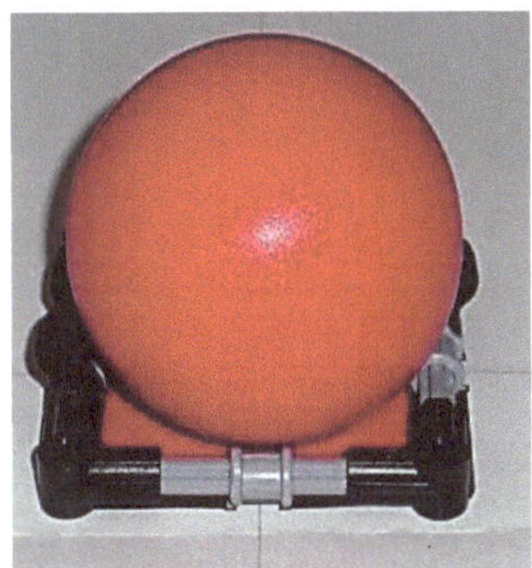

In this mission, you are required to start inside the semicircle "START," go forward and hit the red ball off of it's holder so it rolls outside of the black ring. You are then required to come back to start, and then make a turn around the ball holder and come back to start. Try to do this as quickly as possible, but remember, it needs to work 4 out of 5 times to be considered a successful mission run.

Draw Manipulator:

Program File Name:

Mission Path:

Data Collection:

____

____

____

____

Wall Guides and Wall Following

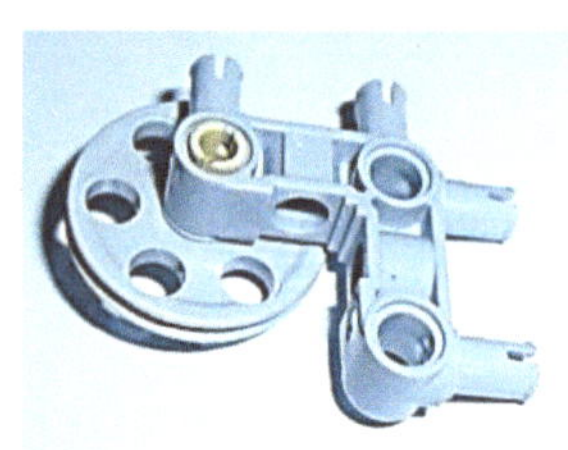

Sometimes when you have your robot go long distances, your robot may run into the walls. We can actually use the walls to our advantage to get places by building a wall guide manipulator. This is a form of passive correction, just like the squaring up exercise.

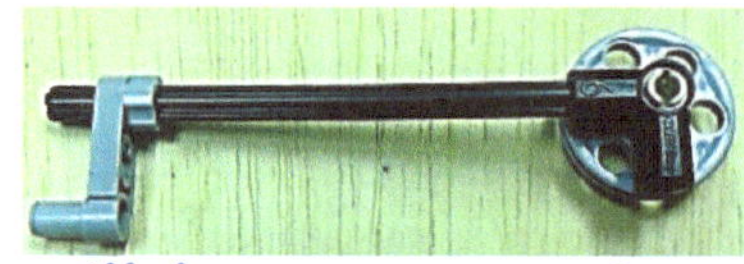

Here is one being used. You will have to try it out in several attachment points on the robot to get it to work correctly. You may even need two – one towards the side/front and the other at the side/back. Build and try the robot event below!

Event #1: Rodeo Hair Pin Race

In this event, your robot must start backwards behind the start line, make a turn, 'run' down the arena, make a turn around the obstacle, and race back to the finish line. The fastest time wins. The obstacle must not be touched or moved by competition robot. (This race is called the hairpin because in a rodeo, the robot "horse" runs in a path around the barrel that resembles a hairpin).

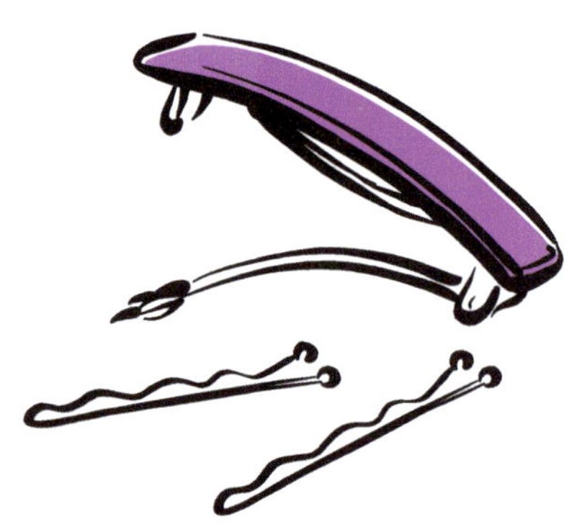

Program File Name: ____________________

Chute #1

Run Try	1	2	3
Time			

Official Time	
Place:	
Points	

UNIT 5
MANIPULATORS & STRUCTUAL DESIGN

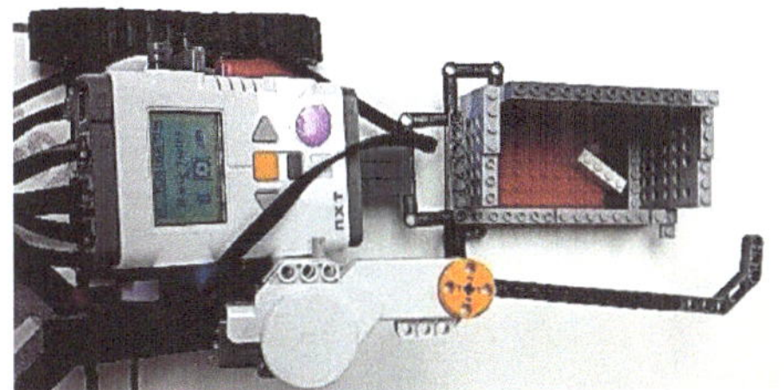

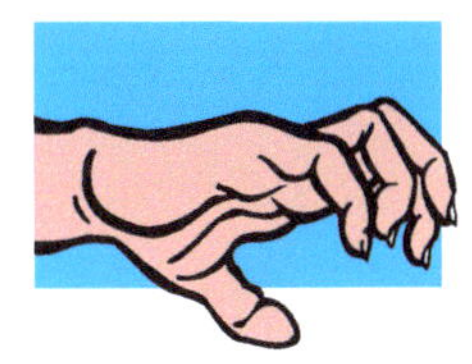

In robotics, a manipulator is a device used by the robot to manipulate materials. A manipulator is usually an arm-like mechanism on a robotic system.

There are many types of materials that are needed to be controlled by robots. The materials could be radioactive, biohazardous, very large or small, or in inaccessible places.

What's the difference?

Passive manipulators lack a motor

Active manipulators are motorized or pneumatically operated.

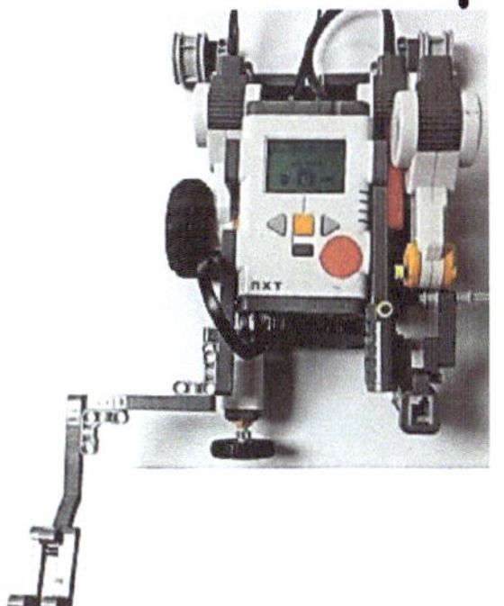

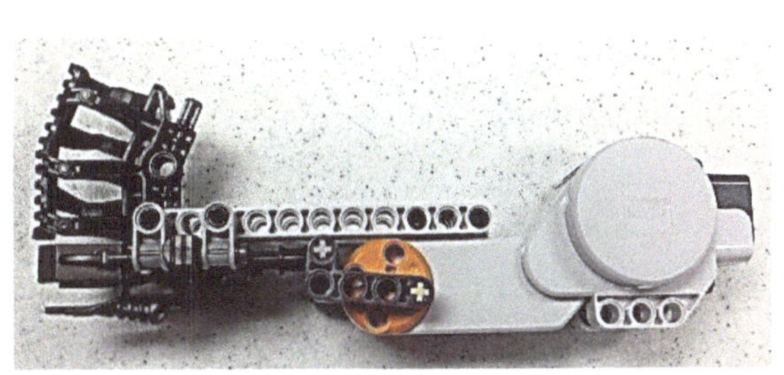

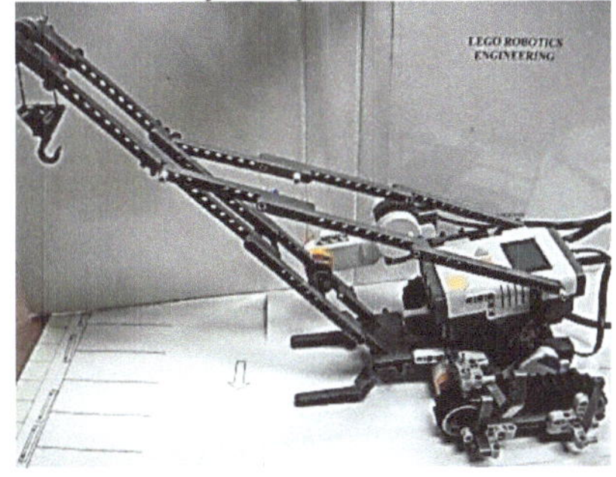

Structural Design

"Structures, Structures, Structures, Yeah!"

An important aspect of robotics engineering is building structural design. Some robots need to be made with parts that are solid and rigid, while other parts need to be flexible and movable. A good understanding of structural design of our LEGO® parts and pieces will help you to build robots and manipulators that do the job or task mission.

Before we get to structures, you need to understand LEGO® geometry.

Expressing Sizes and Units

I have stated in class that each and every LEGO® piece has a specific name. The two main categories of the pieces are the *studded* (with bumps) pieces, and the *studless* (without bumps) pieces. You express the different sizes of the pieces by the studs, also called the units. Builders express the pieces by the length, width, and height. Here is a drawing of a regular old studded 2 X 4 brick:

Note that 3 plates = the height of 1 brick:

Making Beams Extended

Our longest beams are the 15 beam, which has 15 holes. Sometimes you may need to extend this beam on a manipulator or other part of the robot. You need to use connector pegs, and if you want it to be solid and secure, you need at least 2. If you connect them with 1 connector peg, the structure will pivot at that point:

Connecting Beams Cont.

By attaching beams with 2 connector pegs, you cause the two beams to be next to one another and not in a line. You can remedy this by attaching a 3rd and 4th beam with 4 long connector pegs:

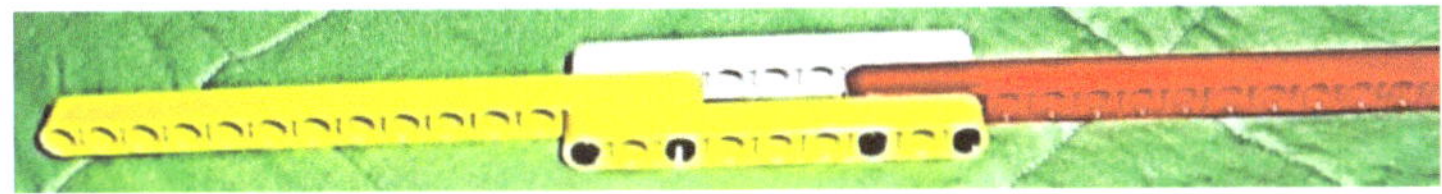

You can also make beams wider by using this same method:

Connecting Parallel Beams:

Put beams in a parallel position. Let's look at two different forces that act upon these beams:

A force is a push or a pull. Tension is a force that lengthens or stretches out an object. Compression shortens or compacts an object. See how tension and compression act upon two parallel beams, and DRAW with arrows the forces that can happen upon these beams!!!

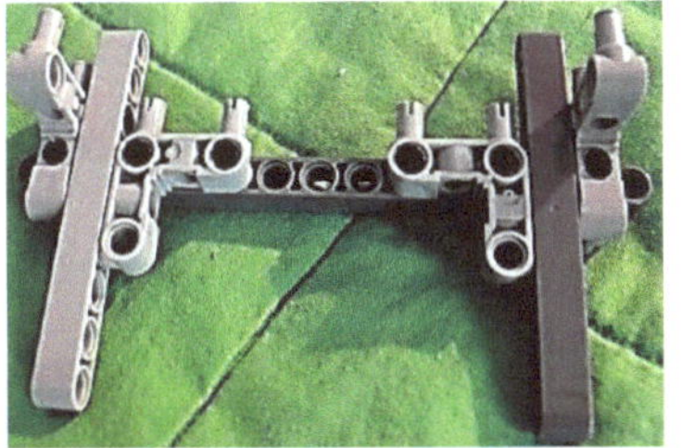

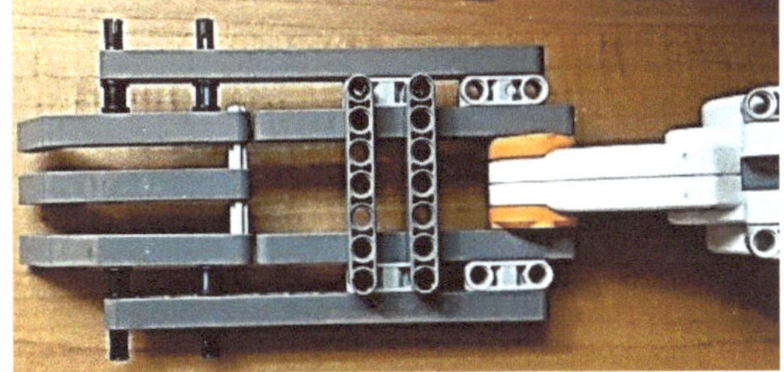

Sturdy Structures

Try building the models below with connector pegs. What shapes are the strongest? Using the LEGO®s, build a square and a triangle using beams and connector pegs. Once built, push the sides together. What happens? Do they have stability?

How can we make a structure rigid and stable? By changing the ______________________

You can easily see that the square structure collapses and slides, and the triangle remains solid.

To make the square strong, you must use a brace by making a diagonal. This really just makes 2 triangles!

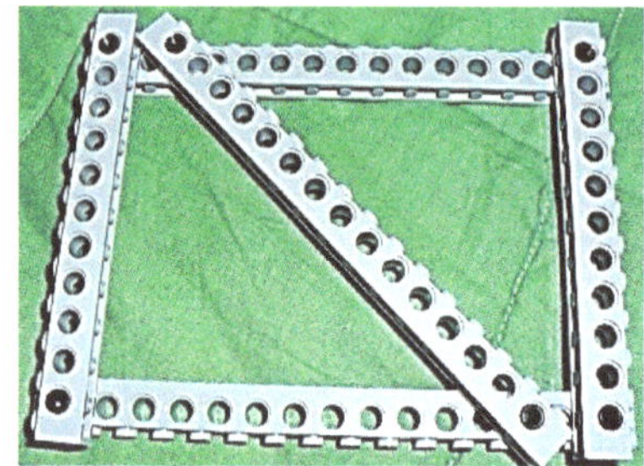

The triangle is the strongest shape. Triangles are used to make a very strong form called a ***truss***. They are used all the time in building structures and bridges. Look for trusses on bridges especially when you travel through a big city on a major interstate or highway.

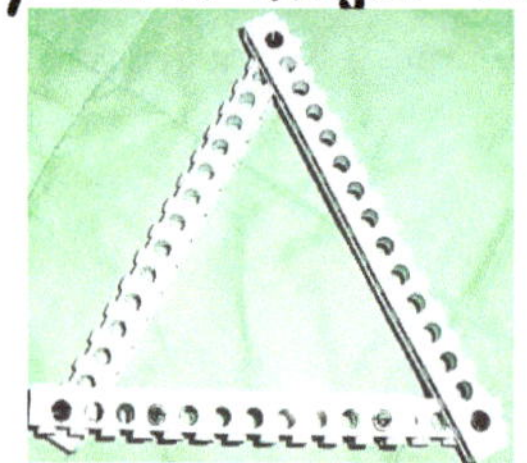

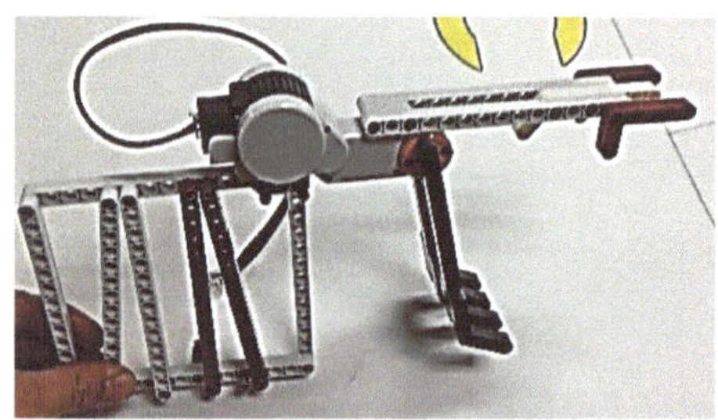

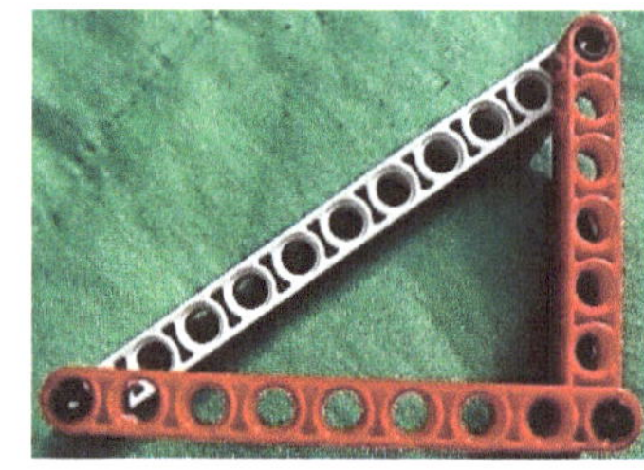

Moving Structures

Some parts of the robot you want to be movable. Here are a few ways that you can connect beams so that they move and extend:

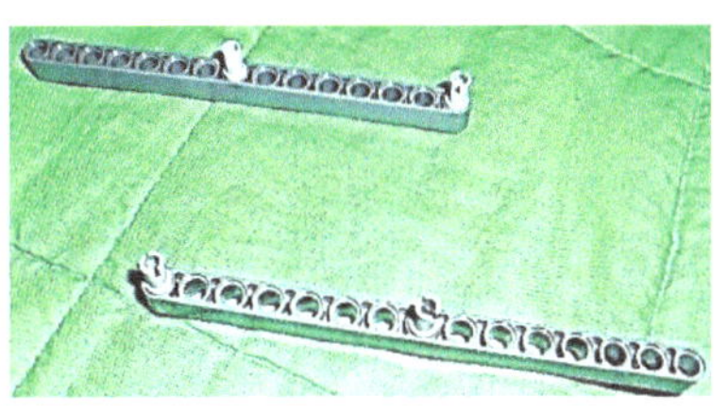

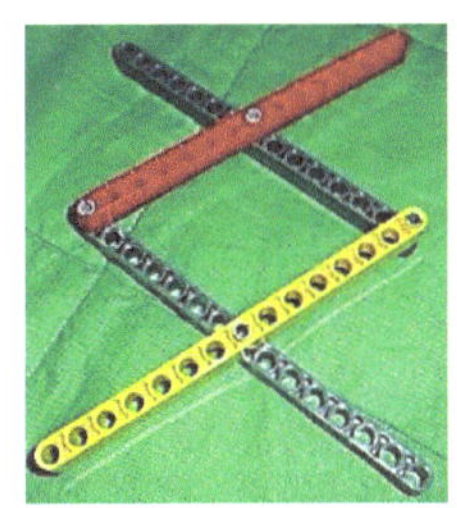

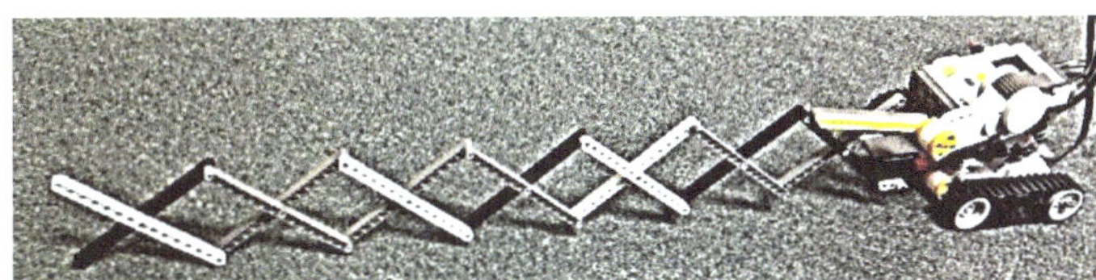

You can keep adding on beams to this scissor-arm style extender:

Building Jigs

When running missions, students often forget to start their robot from the same position each time. It becomes especially difficult to start exactly from the same point when you are starting on an angle in base.

This can be remedied with the use of a jig. A jig is made out of LEGOs, usually out of beams. It forms the angle you need against the wall or edge of the map, and then you line up your robot to the jig before you start.

Here is a picture of a jig of an unusual starting position. It will help the robot pilot to set the robot in the appropriate position each and every time he or she does a mission run.

Remember, the robot and manipulator have to fit inside base and not hang out into the mission field.

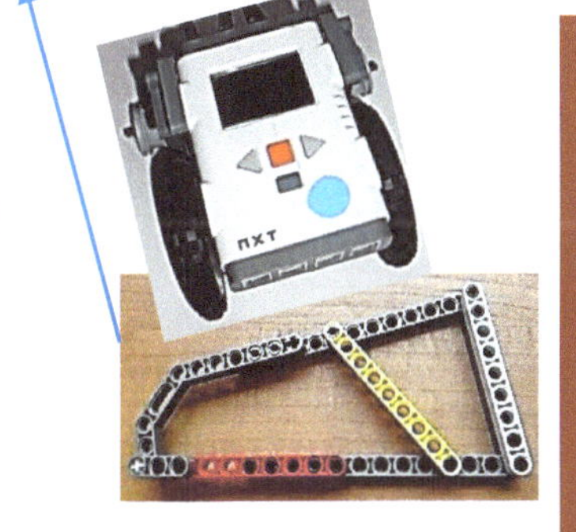

APPLY: Draw a picture of the jig you used in a mission run.

Passive Manipulators

•A passive attachment manipulator is one that is attached to the base robot and is controlled mechanically or by maneuvers rather than with powered motors. There are many different types of passive manipulators, such as:

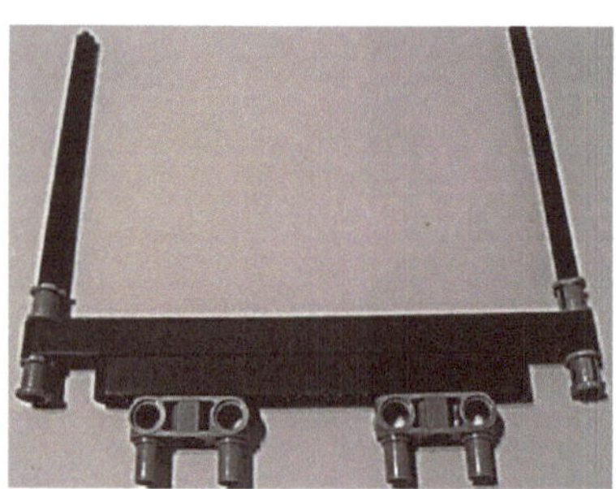

- **Pushing manipulators**
- **Hook/Fork manipulators**
- **CG manipulators**
- **Lever manipulators**
- **Collecting manipulators**
- **Dumping manipulators**

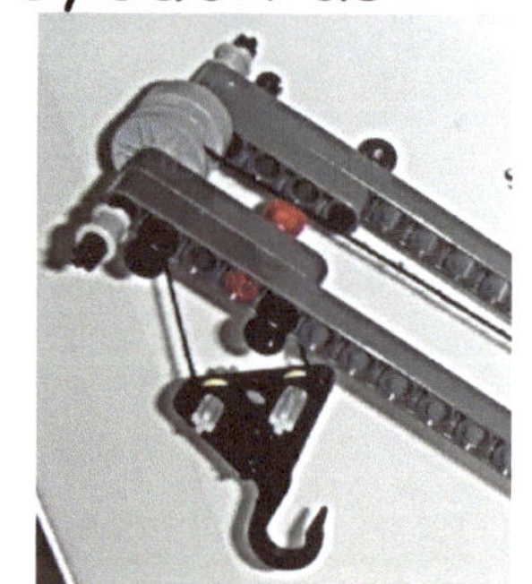

Pushing Manipulators

The pushing manipulator is great for making deliveries and retrievals across the mission field. They can be bumpers, plows, and delivery boxes. The bumper can be a flat wall of beams, and sometimes you can add sides so that the objects do not slip away when making a turn.

Plow Manipulator Exercise:

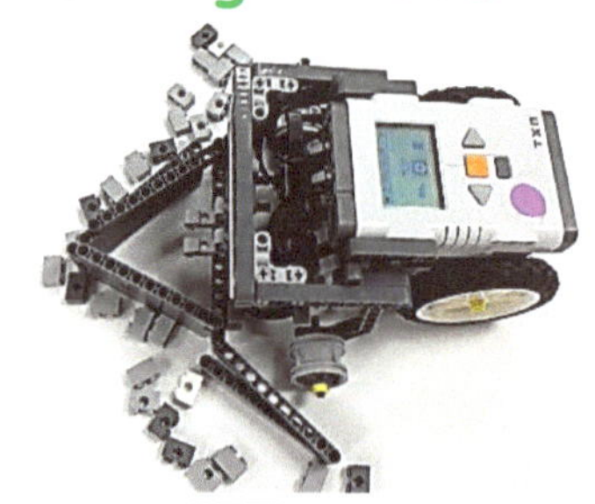

Sometimes you may need to clear through a path of rubble or rough terrain, so a plow works great for these. Design and build a plow manipulator for this mission (a cow catcher design works well).

Draw a sketch of your manipulator:

Center of Gravity (CG)
Manipulators

The Center of Gravity (CG):

What is the center of gravity? It is a point in an object where the rest of the object is balanced around when pulled upon by the force of gravity.

Every object has a center of gravity. Find the center of gravity of a pencil or a spoon by balancing it on one finger. When you get it balanced, you have found the center of gravity of the pencil. TRY IT!

You have a center of gravity too. When you stand up, your CG is in line with your body, But if you bend over at the waist, your CG moves forward, and if you lean too far forward, you may have to take a step or put out your arms so that you do not fall!

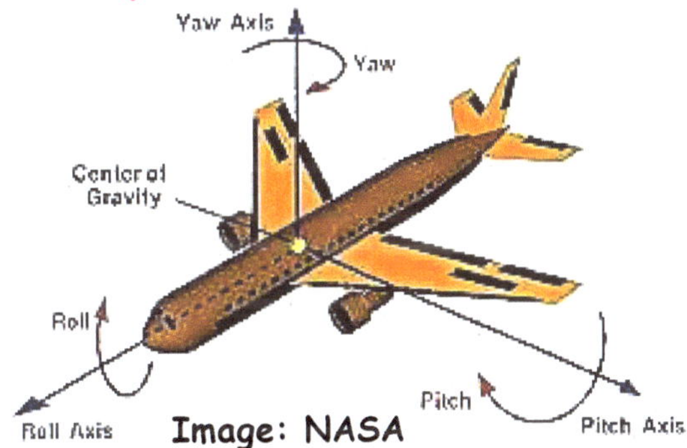

Image: NASA

- We use this idea when we want the manipulator to fall forward in a task. You have the manipulator attached via a fulcrum, and placed back over the base robot so that its CG is over the robot. The momentum of the manipulator and the robot are the same.
- Then, when the robot stops, the momentum of the manipulator is still moving forward, so its CG has shifted forward. The CG keeps moving forward away from the base robot, causing the manipulator to fall.

Center of Gravity Manipulator Mechanics

Here are the mechanics of a CG manipulator:

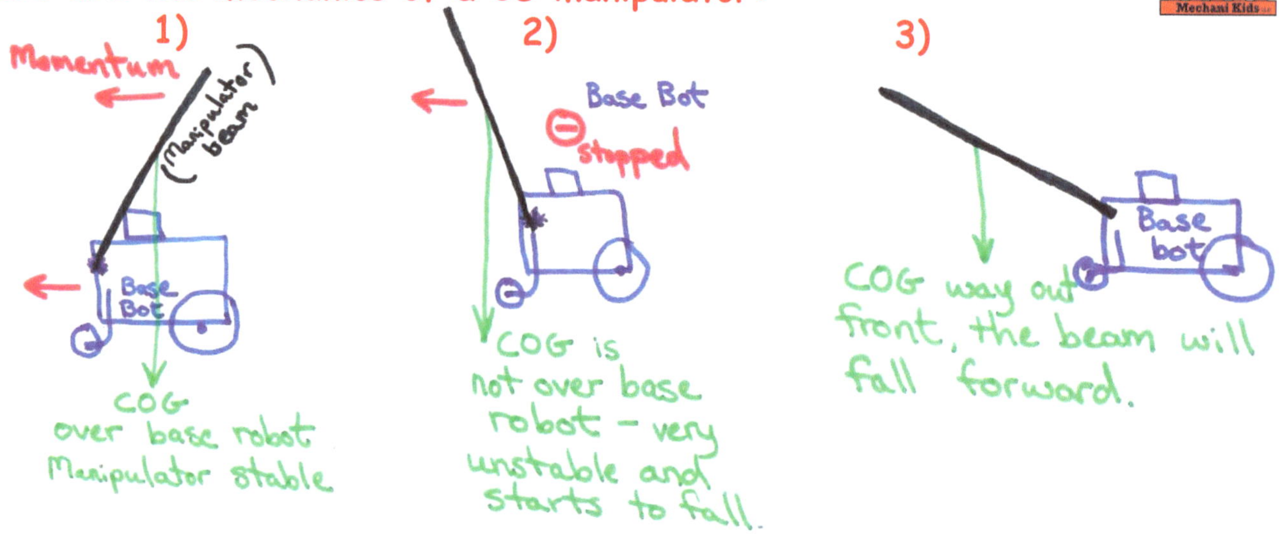

1) Momentum (Manipulator beam) Base Bot — COG over base robot Manipulator stable

2) Base Bot stopped — COG is not over base robot - very unstable and starts to fall.

3) Base bot — COG way out front, the beam will fall forward.

The trick with the CG manipulator is to place the CG of the manipulator as close as possible to the edge of the base robot but yet still completely in balance. This is so that when the robot stops, the CG will need to shift only a little bit to work.

EXERCISE: Lever CG Movement: Have the robot go forward 2 rotations and then make the manipulator fall. How would you do this?

START

You can also do something called a **"Trip step."** This is a quick back and forth movement to help throw the CG of the manipulator forward. It is made by setting two Move blocks, one forward, the other backward, and each is programmed for only around 0.2 SECONDS. What power settings should you use? Make sure to have it set to brake!

EXERCISE: Trip Step Ex: Have the robot go forward 2 rotations and then make the manipulator fall by doing a "trip step." Make sure to come back to base!

START

Lever Manipulators

The Lever Manipulator:

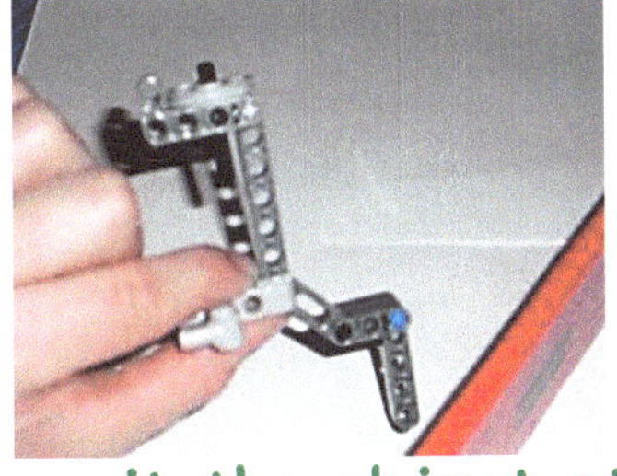

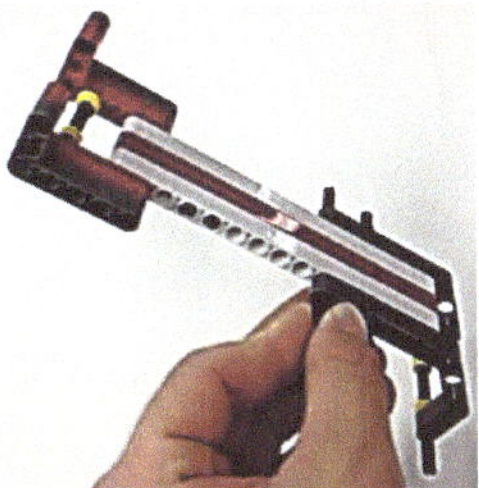

•The lever manipulator is a great way to deliver an object by holding it when you can't push it out in a task mission.

•It uses the simple machine (lever) to deposit the object at the right time. It works great for delivering to a basket, onto something, or to trigger an obstacle to do something out in the mission field.

You can build one easily and quickly by using:

•2 2x4 gray L beam
•A 1x11.5 double bent beam
•A 3x5 gray L beam
•Blue axle pin
•Black connector pin
•Axle 3
•A Hassen pin

The way it works:

This manipulator uses a 1st class lever. A first class lever has the effort between the fulcrum (pivot) and the load. The load is the object you are carrying out to be delivered, and the effort comes from bumping into the target area (a wall or obstacle out in the mission field).

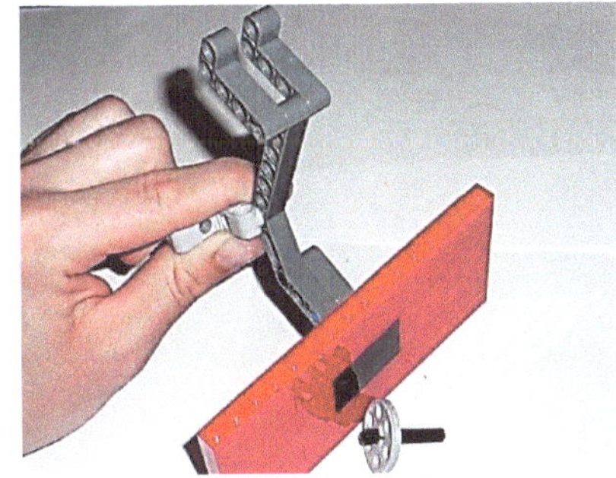

Like all manipulators, it should be made as stable and simple as possible. It is also best to make the lever more upright so that the object you are taking only drops into the target, instead of being flung in the general direction of the target.

REMEMBER - when designing and building manipulators, your goal is to build something that is very reliable, and that repeats the movement again and again. You do not want the manipulator moving a different way each time you test it.

Forks

•Forked manipulators are also great for retrieving target objects.

•Forks have tines on them that make it easy to slip on the target object so that the target object doesn't move. Then the catch is made when the robot makes a turn or forward movement. Forks are great because you can make lots of tines, so that you have a higher probability of getting the object!

Catch the Loop Retrieval Exercise:

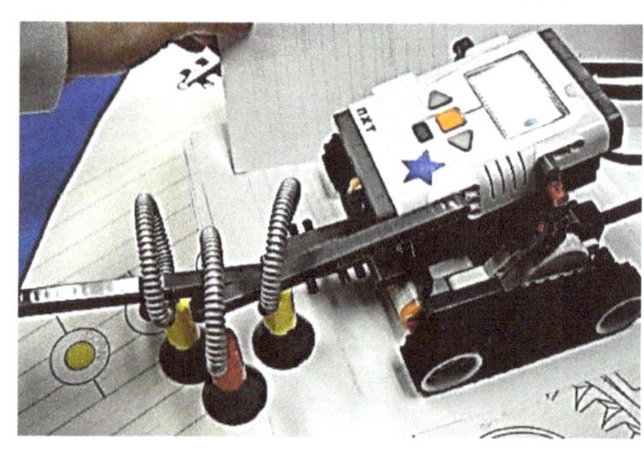

Go out on the mission field and retrieve the loop, and bring it back to base using a pronged fork manipulator.

Drawing of Path:

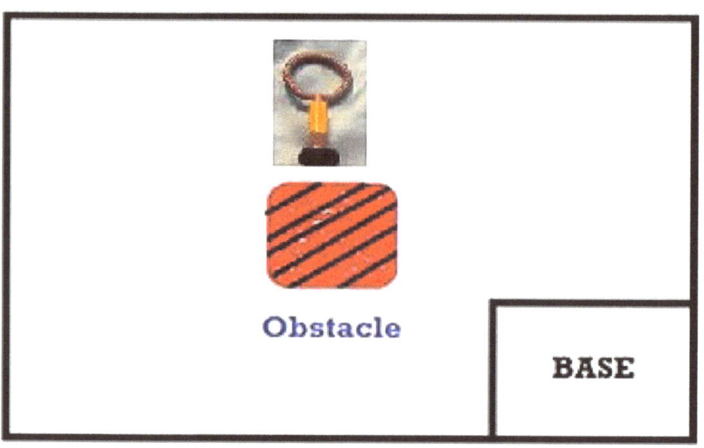

Drawing of Manipulator:

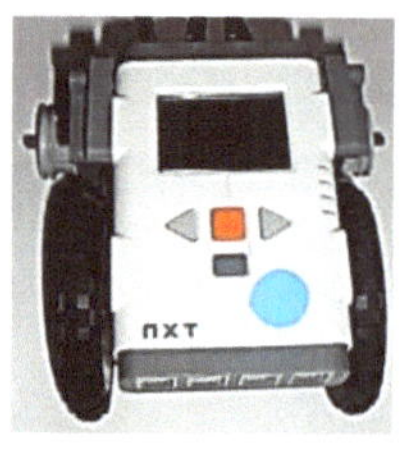

Write Program File Name:______________________________

Fish Hook Manipulators

•Fish hook manipulators are great for retrieving target objects that are not resting on the ground in the mission field.

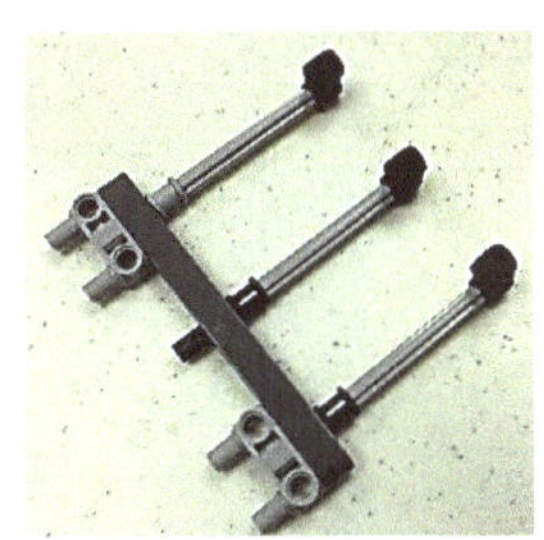

•Many fish hook manipulators are passive attachments that are controlled by a good program. The program needs to contain movements of the robot that may go past the target object, and then with a turn or backing up movement, "catch" the target object, and drag / pull it to base or wherever it needs to go.

Simple Fish Hook Manipulator

This type of fish hook has a barb on it to catch the object and then have it not fall off the hook while transporting.

Fish Hook Retrieval Exercise:

Go out on the mission field and retrieve the "I" brick beam, and bring it back to base using a fish hook manipulator.

Drawing of Path:

Drawing of Manipulator:

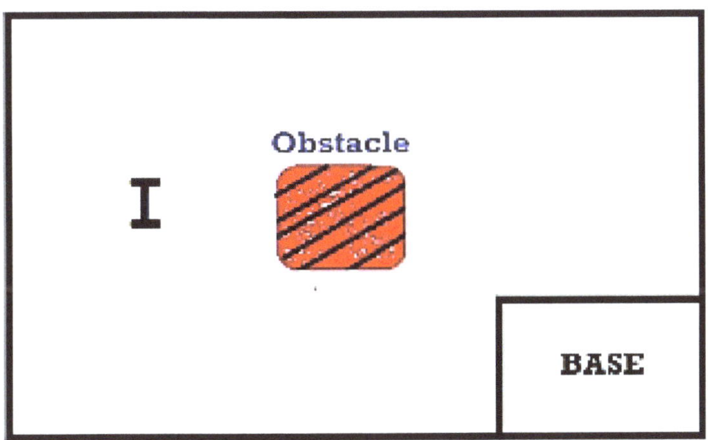

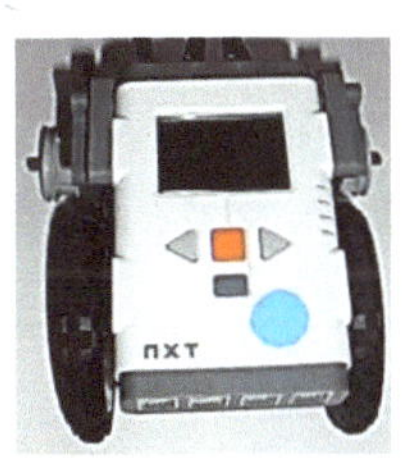

Write Program File Name:______________________________

Roller Bot

•Meet our new robot build: Roller Bot!

•Roller Bot is a totally different type of robot than our Wheel Bot. It is not a skid bot, but instead it uses a caster wheel for stability instead of two wheels which lack tires.

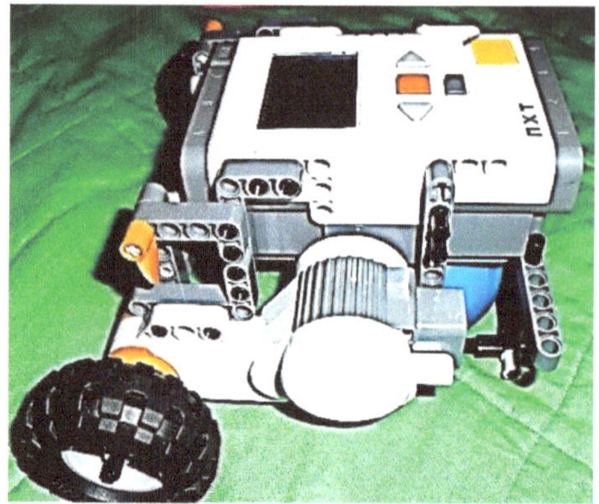

The back wheel is a caster wheel – which means it swivels and turns easily as it follows behind the robot. Caster wheels come in all shapes and sizes. The one we are going to use today for the castor wheel is actually one of our LEGO® balls! Using the ball as a castor wheel is great because the ball helps to make the robot make nice turns and backwards movements.

This robot is also:

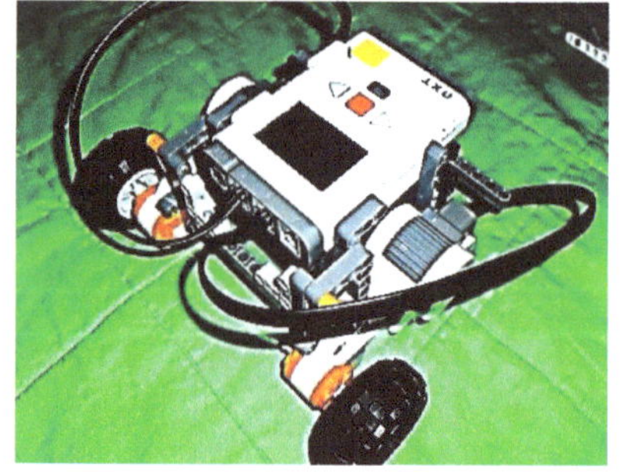

•Compact in size
•Symmetrical in design
•Uses a differential drive system
•Uses B and C motors for locomotion
•Has a shifted CG that is more towards the tire wheels.
•Has small balloon tires for better traction.

Build and attach the beam grills for the robot so that you have many attachment points for manipulators and sensors. Remember to plug in the cables with the letter port on the IB matching the same side of the robot's wheels.

Casters & Skids

There are many different types of skid wheels that you can use on the robot. You need to figure which type of skid works best for you, keeping in mind a few different key points about your robot:

1. The size of the robot - is the robot spread out over a large area, or is it compact?
2. The weight of the robot - is it heavy or light?
3. The terrain or mission surface - smooth, carpeted, or rubbery?
4. Where is the center of gravity of the robot?

Find the answer to these questions on your robot.

You may need to use a different LEGO® element for your skids or caster, the goal is to have the most stability with the least friction.

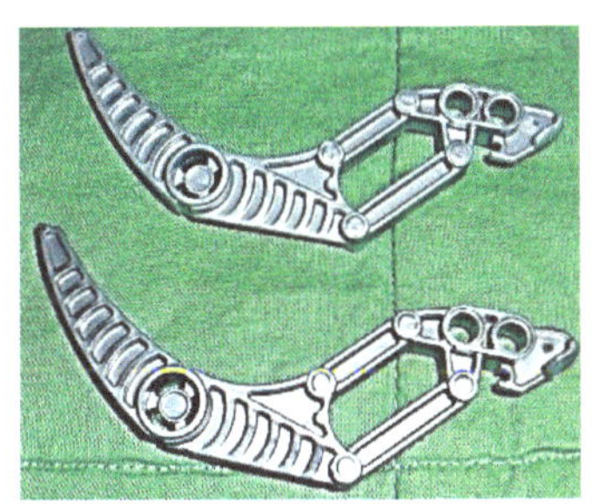

The claws are great skids, and are very light and easy to attach with their double holes. They also look cool!

The red or blue balls are great to make a caster wheel that doesn't jog when going in reverse. Think of an old-fashioned computer mouse. They must have a cage around them, which can sometimes cause friction.

The inverted round tile plates are great for being very friction-reducing. They work great on medium weight robots.

These small wheel hubs are great and also roll when going forward and back.

Dumping Manipulators

•Dumping manipulators are also great for making deliveries to target objects. With a good dumping manipulator, you can make "Speedy Deliveries!"

•The manipulator on the above left is a very complex dumping manipulator, so let's start with something simple: a lever.

•Like a dump truck, the load is placed in a bed and then when a trigger is pressed, the load is dumped off the bed. You can use a trigger as a lever to cause the back end of the bed to rise and tilt, and pivoting around the front of the bed, causing the load to fall.

Tree Dump.rbt: Plant some trees. Take baby seedlings and go out to the nature preserve for planting. Use a dumping manipulator for this mission, and return to base. Dump the trees over the fence:

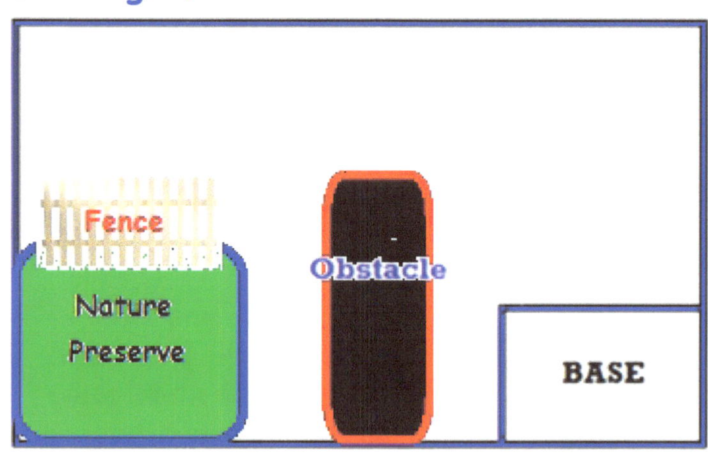

Drawing of Manipulator:

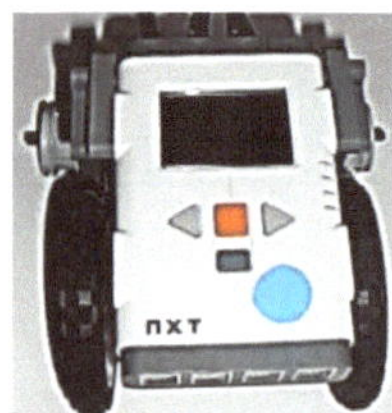

Write Program File Name:____________________________

Collecting Manipulators

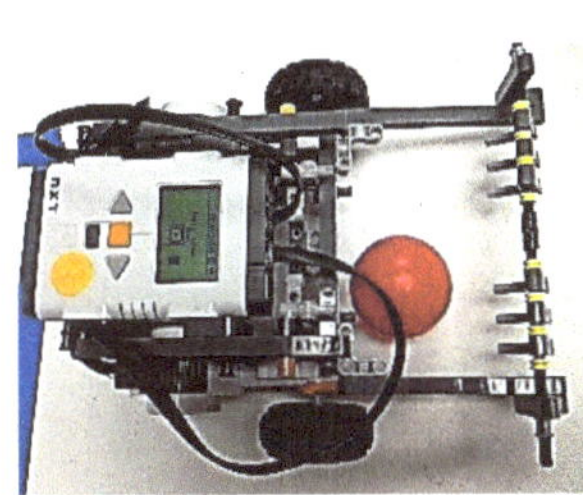

•Sometimes your mission may require that you retrieve an object that may be difficult to catch hold of with a hook or a forked manipulator. A good example of this is a ball. Balls tend to get rolling and can easily get away from your robot.

•A collecting box is a passive manipulator that has a swinging door on it which opens one way to catch the object, like an animal trap. Once the object is inside, it cannot get out of the door.

The peg attachments on top allow the door to only swing to the inside.

Ball Collect.rbt: Go out and catch the 2 balls on mission field using a swinging door colleting box. Return to base.

Drawing of Path:

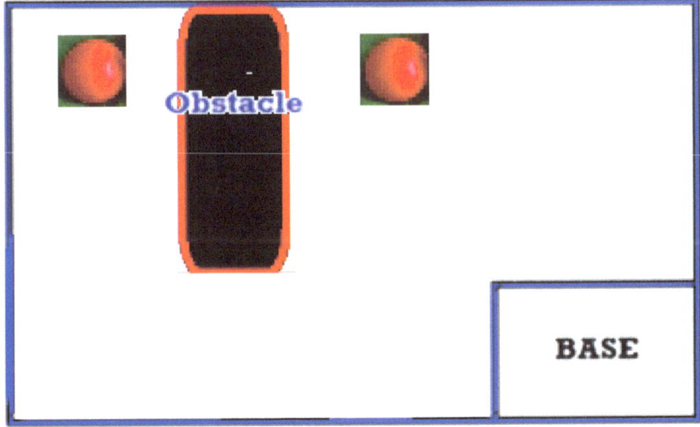

Drawing of Manipulator:

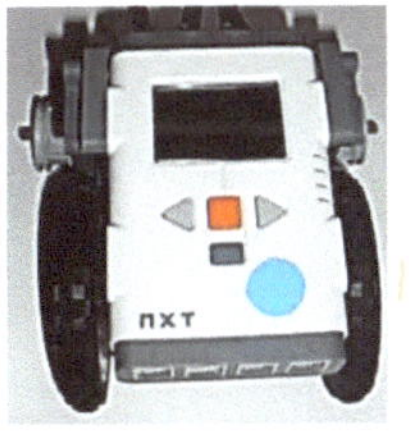

Write Program File Name:______________________________

Sweeping Manipulators

•Another way to collect items out in missions is to build a sweeper machine. This machine has bent beams used like a vacuum cleaner attachment to catch the items. The "bristles" turn due to a gear train that takes energy from moving across the map and turns the sweeper. Make sure that the bristles turn the correct direction!

The 40 toothed spur gear is an idler gear. What purpose does it serve?________________

____

____

Sweeping Duty.rbt: Go out and retrieve the 2 balls on mission field using a vacuum sweeper colleting box. Return to base.

Drawing of Path:

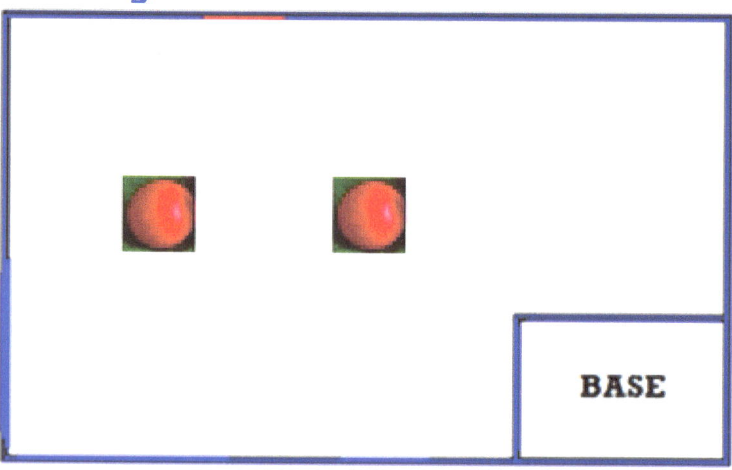

Drawing of Manipulator:

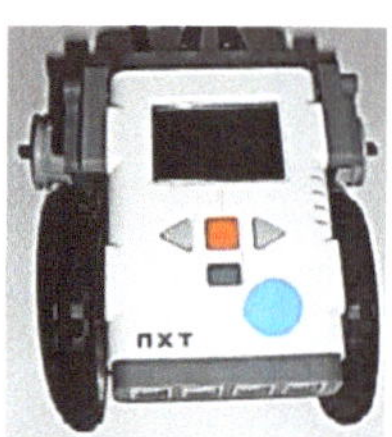

Write Program File Name:______________________________

Motorized Manipulators

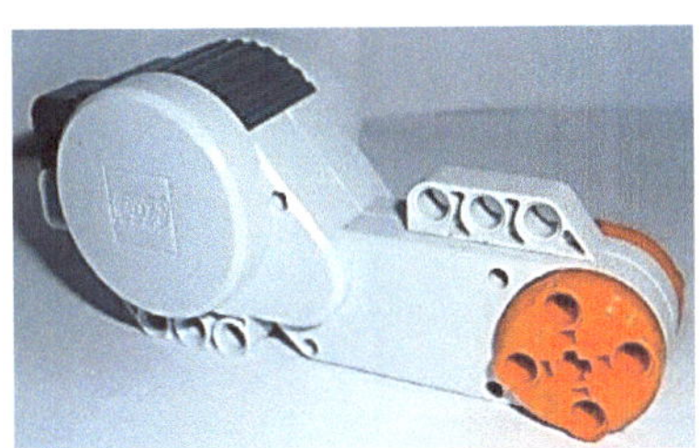

•Motorized manipulators are a type of powered manipulators. They are different than our passive manipulators (Center of Gravity (CG), Plows, 1st Class Levers, Hooks, Forks, Sweepers, Collecting Boxes, etc, because they are moved by a SERVO motor!

•This is where Port A comes in! Finally, this port can be used to control Motor A.

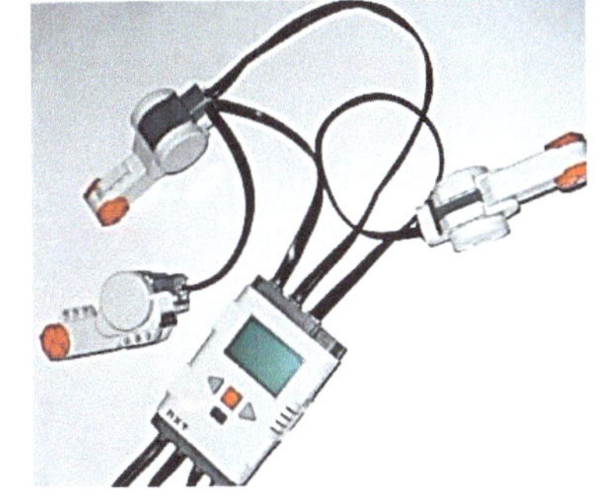

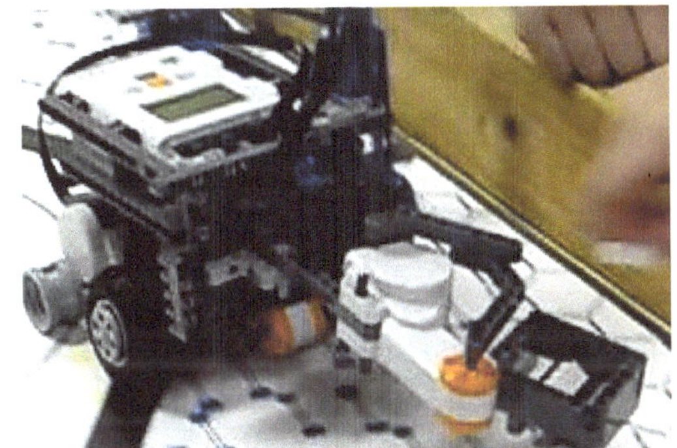

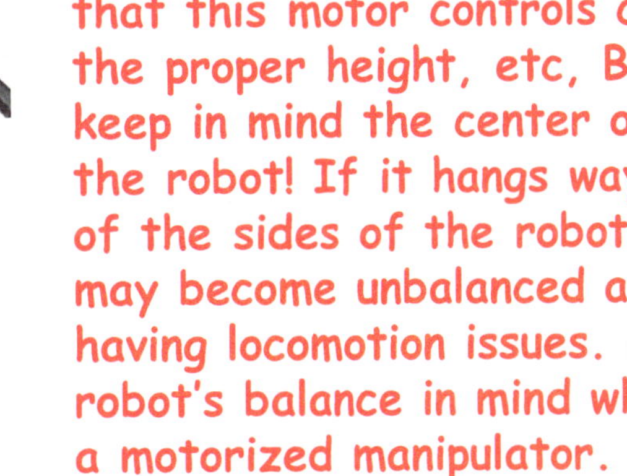

•The place where you attach your 3rd motor takes a little bit of thought. You want the motor so that the manipulator that this motor controls can move at the proper height, etc, BUT you must keep in mind the center of gravity of the robot! If it hangs way out to one of the sides of the robot, your robot may become unbalanced and start having locomotion issues. Keep the robot's balance in mind when attaching a motorized manipulator.

Programming Motorized Manipulators

•When using a motorized manipulator, you program this servo motor just like you would for making the robot move. This time, you will be using MOTOR A. You can use a Move block or a Motor block. If Motor A is being used by itself, I would recommend using the Motor blocks when making manipulator movements.

Look at this lever beam attached to Motor A. What duration would you program to make this lever beam move?

Motor Manipulator Exercises

Motor Tap Object.rbt: start with the lever beam upright and then tap the object. Come back to base.

Drawing of Path:

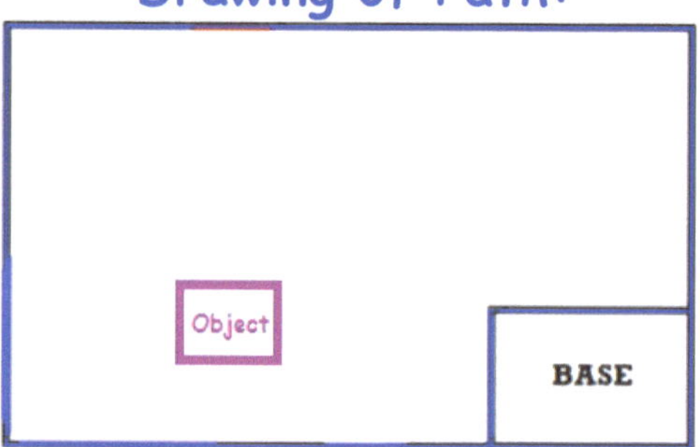

Write Program File Name:

____

Drawing of Manipulator:

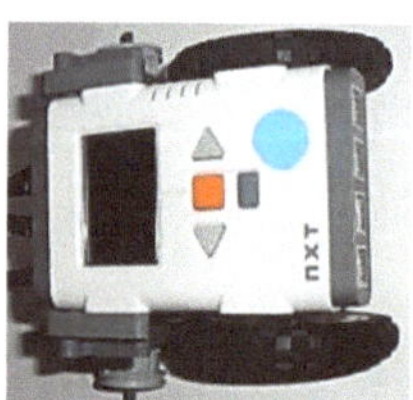

Loop Lift Object.rbt: start with the lever beam low and then lift the object. Bring it back to base.

Drawing of Path:

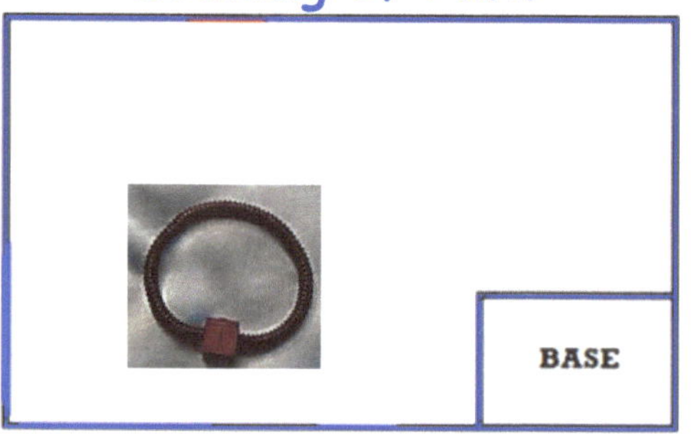

Write Program File Name:

____

Drawing of Manipulator:

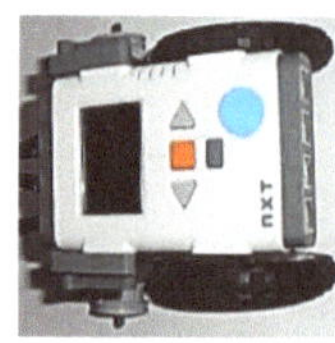

Motorized Claws

A great type of motorized manipulator is the motorized claw. Design and build a motorized manipulator that grabs the target object (empty soda bottle) and comes back to base.

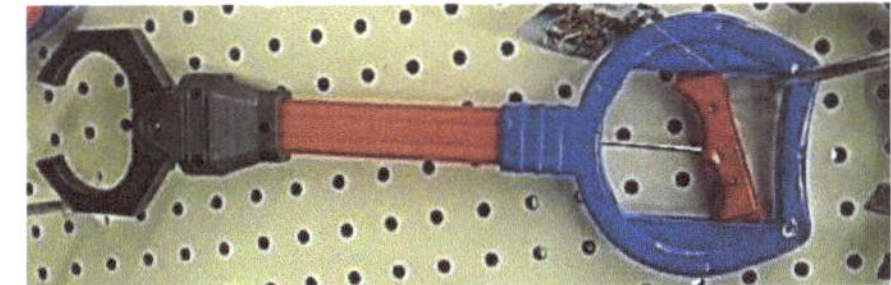

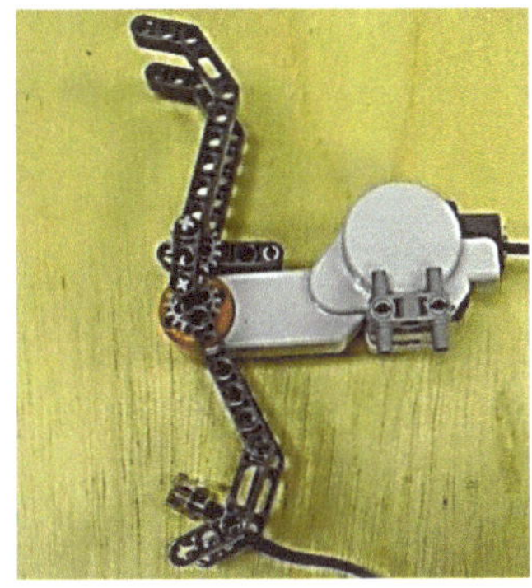

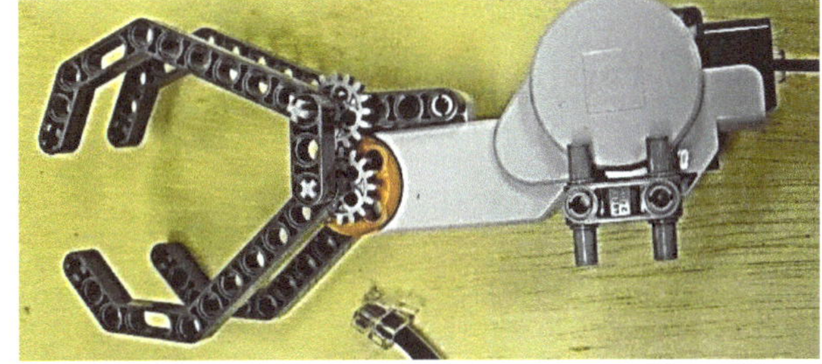

Draw Path:

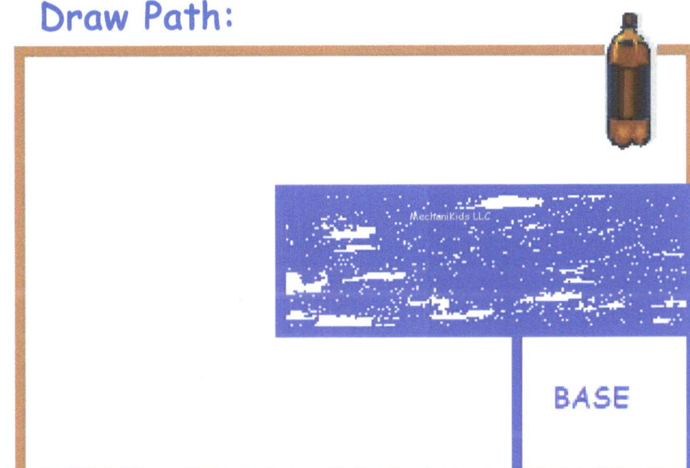

Sketch design of YOUR manipulator claw:

Program File Name:

________________________.rbt

FLOWCHART:

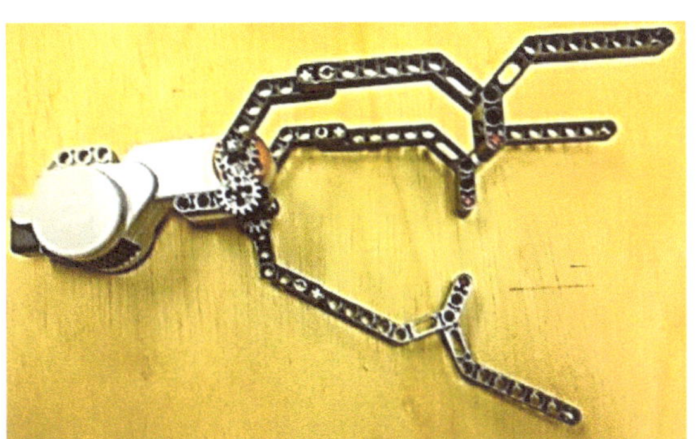

UNIT 6
ROBOT SENSORS & SENSOR PROGRAMMING

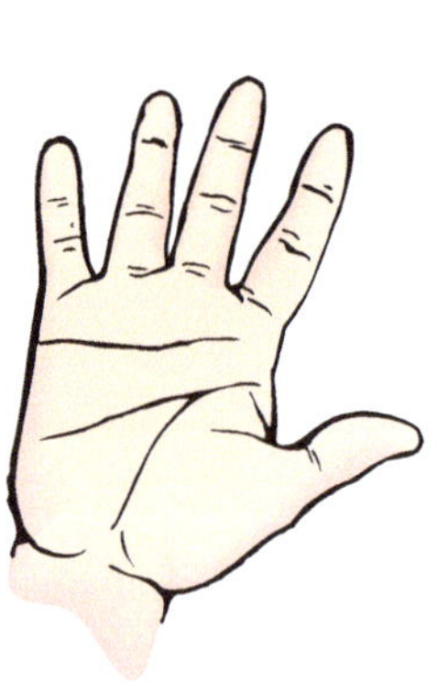

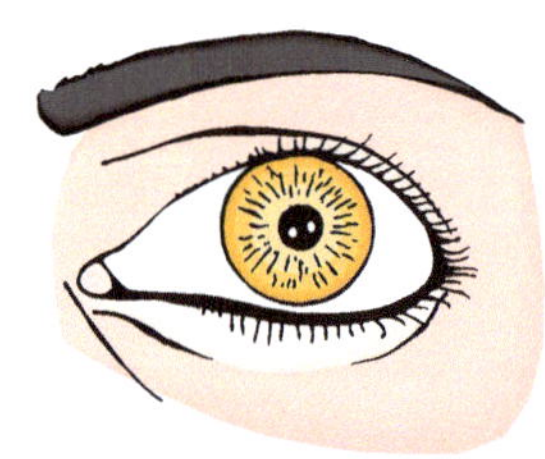

A robot is more than just a bunch of hardware and a "brain." A robot must be able to react to its environment by sensing. Just like we have our 5 senses for our human body to react and gain information from our environment, our robots use sensors to acquire this information.

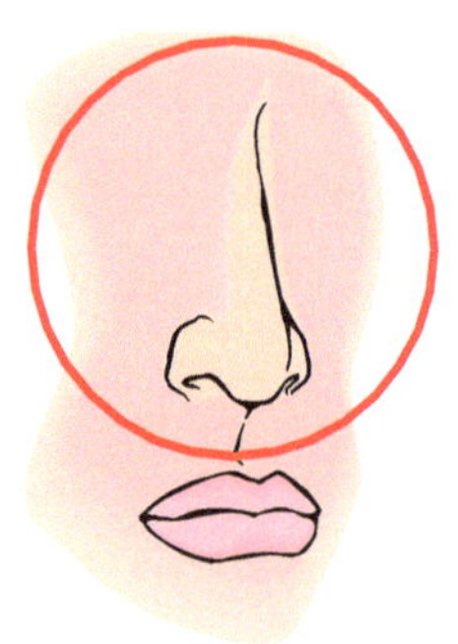

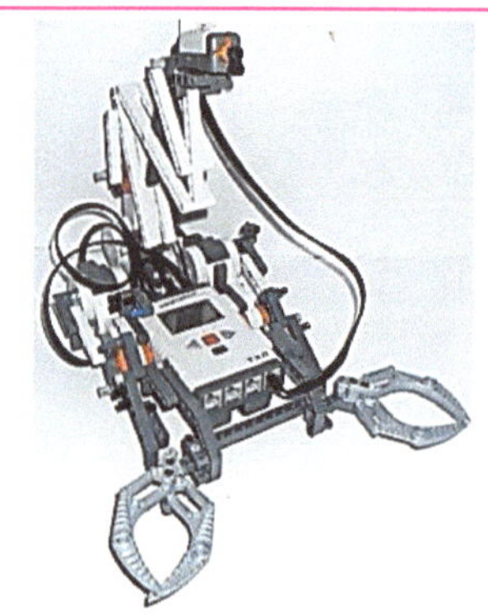

NXT™ Sensors

Although there are many, many sensors for out NXT™, we are going to be using 4 NXT™ sensors on a regular basis:

1. Touch sensor
2. Sound sensor
3. Light sensor
4. Ultrasonic sensor

Touch Sensor

Sound Sensor

•Each type of sensor has a specific Input Port on the IB. Please make sure that you are using the correct sensor with the correct port – this will make programming less complicated!

Light Sensor

Ultrasonic Sensor

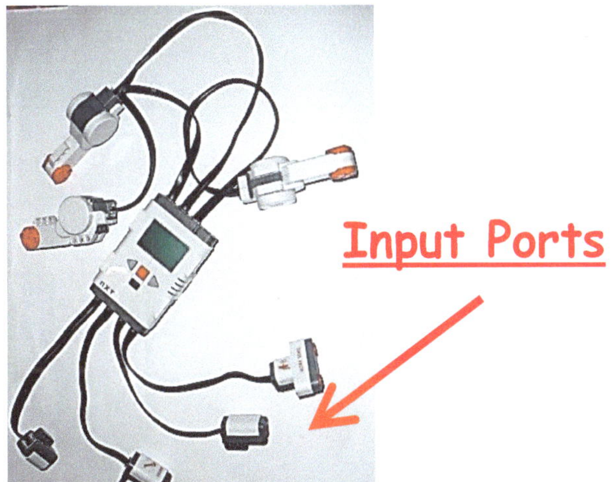

Input Ports

Label the Input Ports:

•Port 1 - ____________________

•Port 2 - ____________________

•Port 3 - ____________________

•Port 4 - ____________________

•We will discuss in detail about the characteristics and uses for each sensor.

Program Flow & Flow Blocks

Program flow is the way and order that the parts of the program run. Just like you can control the flow of a river by changing where it flows, or by stopping it up with a hydroelectric dam, you can control the program flow by using Flow blocks.
The 5 types of flow elements are the sequence beam, the loop block, the wait block, the switch block, and the stop block. These elements control the order in which the program flows, or executes.

Loop Blocks

Loop blocks have a part of the program repeat, and can be set to a certain condition. There are two main parts of the Loop block:

1)The loop body (the "picture" inside)
2) The loop condition (the "frame" of the picture)

You put the Loop block on the sequence beam and then place the other blocks inside the Loop block, which will expand. The blocks inside the Loop block are called the loop body, this is what you are looping.

The Loop condition is what the loop action is set to, determining if it loops or exits the loop and heads to the next part of the program. You can set it on: Count, Forever, Sensor, Time, and Logic.

Setting the Loop block on Forever will cause the program to repeat, well, forever until it is stopped by pressing the dark grey button or using the stop block. Using a sensor condition will loop until a sensor reads a certain threshold and then exits the loop block and runs the rest of the program.

LOOP Block Exercise: Quick Loop Ex.rbt: Have the robot go forward 2 rotations and then stop, back up, and then make a quarter turn. Have the robot first do this infinitely, and then adjust the program to have this movement run 4 times.

The Sequence Beam

You have already been using the sequence beam. It is the LEGO® Technic™ white beam on the workspace that you drop the programming blocks onto controls the running sequence of your program blocks. You must make sure that the blocks that you place are connected or dropped onto the sequence beam. If the block has a gray, fuzzy look to it, then it is not connected and it will not be downloaded to the NXT.™

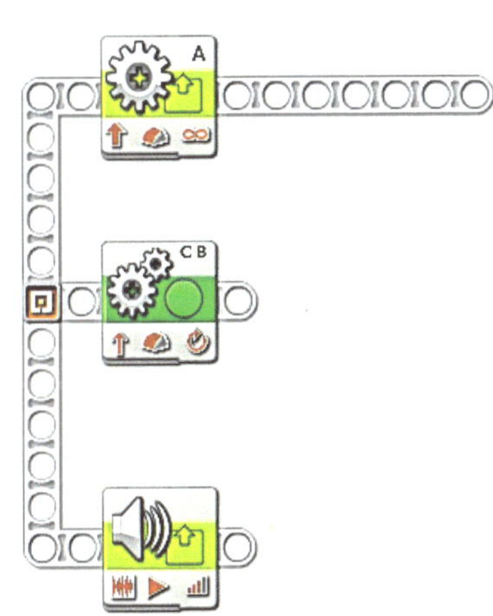

There is also the ability to run more sequence beam as the same time as your main one. This lets the program execute different blocks all at the same time, this ability is called **multitasking**.

An example of this might be a move block running the robot's locomotion, and the motor block engaging motor 'A' to operate a manipulator, all while playing a sound file!

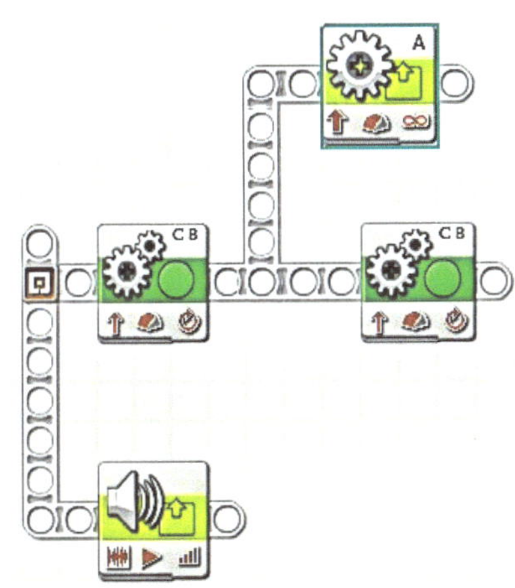

You do this by clicking on the start box at the beginning point of the sequence beam and dragging up and over, and then double clicking. It takes some practice. It can also help to place the programming block on the workspace first and then connecting the beam afterwards.

Touch Sensors

On many mission challenges, it is difficult to get the exact distance to a target spot that is far away down the mission table. One way to help with this is to use a touch sensor.

The Touch sensor is your robot's "fingers" and reacts to touch and release when the button is pressed on the touch sensor. It enables the robot to "feel" and can detect single or multiple button presses, which is sent to the NXT Intelligent Brick. Use a WAIT block to program your touch sensor.

Wait Blocks

The Wait block controls the flow of your program and lets your program run, waiting until a certain requirement is fulfilled, and then once that requirement is fulfilled, it then moves on to the next part of the program. You select or type in a value to set a trigger point so that the program continues when sensor values are below or above it. Use a Wait Until Touch Sensor block to program with the use of a Touch Sensor.

How it works:

The orange tip, when pressed, will send a signal that registers as a "1" to the IB. If not presses, it sends a signal as a zero signal.

There are 3 things you can set the touch sensor for:

1. Pressed
2. Released
3. Bump (Pressed and Released)

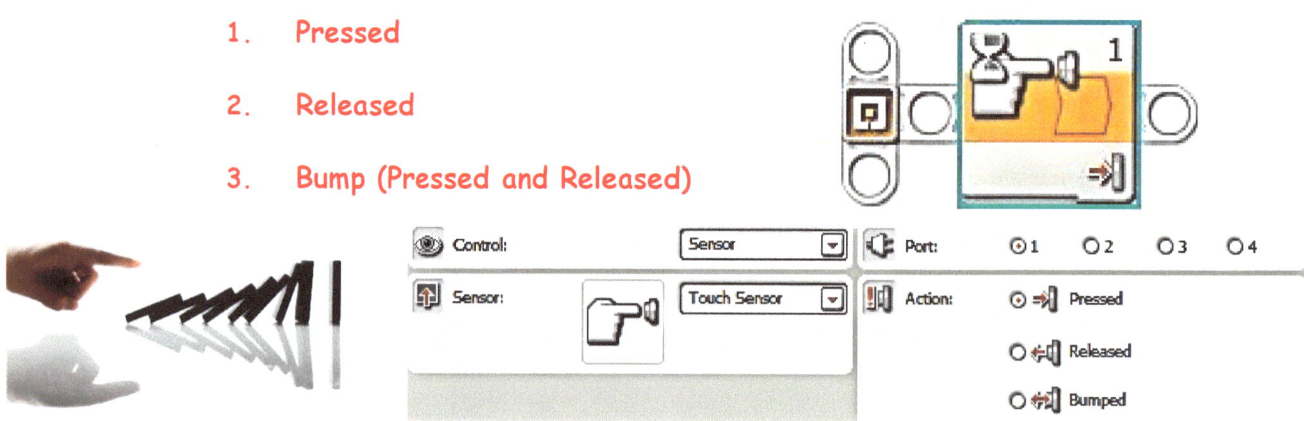

Using Bumpers

A bumper is a type of manipulator that you attach to the touch sensor, or it swings and depresses or lifts off the touch sensor button. The touch sensor has an axle joiner, so most students prefer to use an axle and just stick it in the sensor. The problem is that these may actually hit the obstacle or wall at an angle, causing the sensor depressing button to stick or freeze. I have found that using a type of simple lever that swings and presses into the button works more consistently.

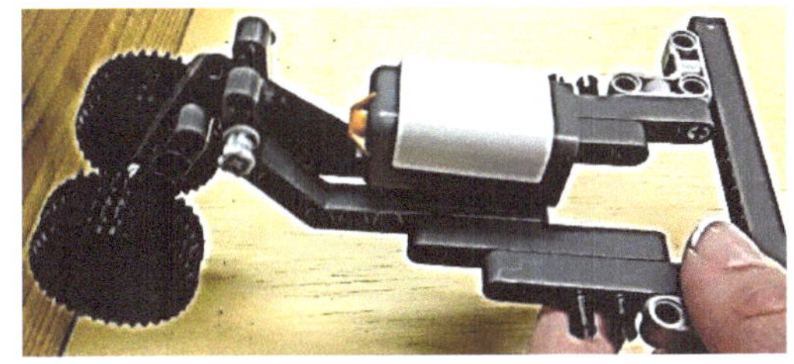

The touch sensor does not just need to be pressed, you can also have a "closed" bumper that is pushing on the sensor constantly and then gets released. It has some way to stay pushed up next to the sensor, and then it can bump into an object, and by using LEVERS, can change the direction of the effort and releasing the button. You will need to change the configuration panel on the Wait Until Touch block to "release."

You can build a manipulator to help trigger the sensor when it touches an object. Build one and connect a cable from the sensor to the IB.

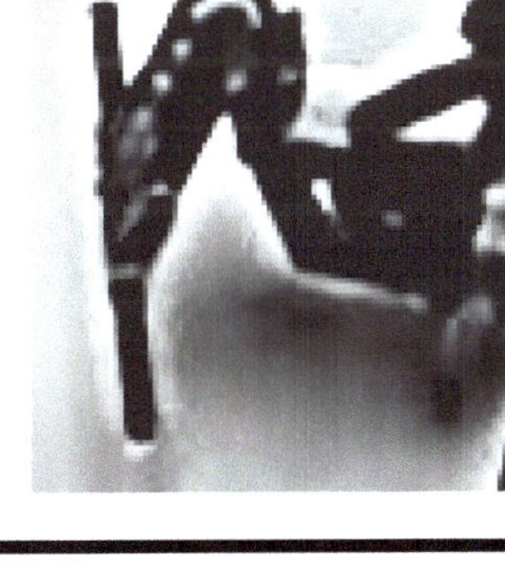

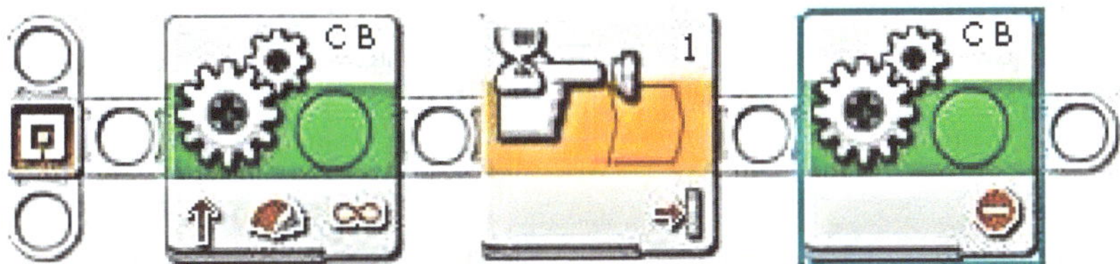

Now for programming, you can use the wait block with the touch sensor setting. Try it on the mission table wall!

Touch Sensor Exercises

·EXERCISE A: Press Go Button Exercise

·Have the robot go forward 1 rotation when you press the touch sensor to start the run. Did you use a Wait block or Loop block?

·EXERCISE B: Touch Sensor Exercise

·Have the robot go forward until the touch sensor comes into contact with a wall and then stops and reverses. Did you use a Wait block or Loop block? Why?

____

____

EXERCISE C: 3 Touch Mission: In this mission, you are required to start inside base, and then hit the 3 different wall check points in a sequential order. You must come back to base.

There are two mission tables, you and your teammate will be at one of them. CIRCLE YOURS!

Draw Path:

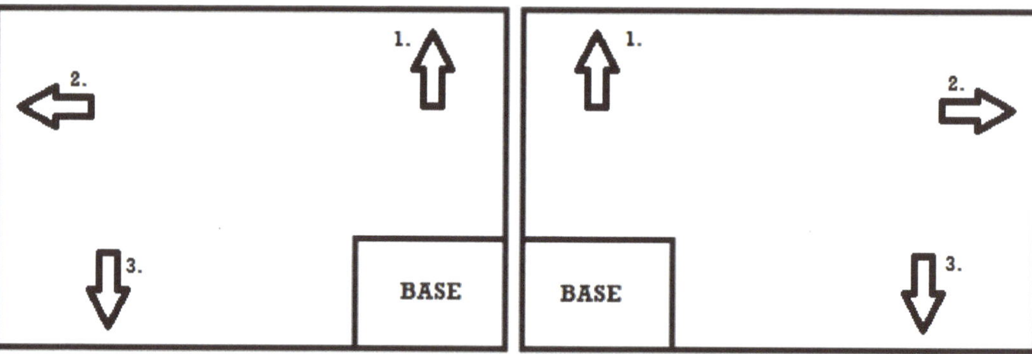

Program File Name: ____________________________

MAKE FLOWCHART:

Sketch of Manipulator / Sensor: *Draw top view and side view, and label parts.

Intro to Light Sensors

•A light sensor is a electronic device that detects light. The light sensor assists in helping your robot to see. It enables your robot to distinguish between light and dark, as well as determine the light intensity in a room or the light intensity of different colors.

•The sensor sends out a beam of light, and then detects and measures the amount of reflected light (the light that comes back) in a percentage on a scale from 0% - 100%. The sensor then sends the percent of reflected light through a signal to the IB.

•A brightly lit room is usually around 90, a ordinary room is about 50, and a dark room may be around 30.

•The extra light in the room can thus affect the light sensor. This extra light in the room is called the ambient light. Each room is different (bright lamps, windows, etc.), and you need to measure the light in the room each time you program the light sensors. You get this measurement by going to the View mode, select reflected light, and then go to Port 3. The percentage of reflected light will show on the display.

•When using light sensors, you need to make sure that the sensor is pointing down and is as close as possible to the surface. You will get an inadequate reading due to "lost" light if the sensor is high or on an angle.

Reflected light

Sent light

Surface

To program the light sensor on the robot, we need to find the threshold. The threshold is a specific point or level. The way we find this threshold is by finding the average. The average (for 2 measurements) is the middle point; it is (measurement 1 + measurement 2) ÷ 2. The average is the value that is equally distant from the measurements of the white and black areas.

Light Sensor Exercises

Measurement Exercise: Let's try finding the average for ourselves: Go to View and select reflected light. Then go to Port 3 and place light sensor on the white area and then black area.

Measurement of white area:____
Measurement of black area:____

Now calculate the average:
(measurement 1+measurement 2) ÷ 2 =
(_____ + _____) ÷ 2 = _______
This is the number you program the light sensor on the computer.

EXERCISE : Light Sensor Test
Have the robot go forward unlimited until it runs over the black paper. Program it to stop when it detects the black paper. Did it work?
Program: TRY both of these programs:

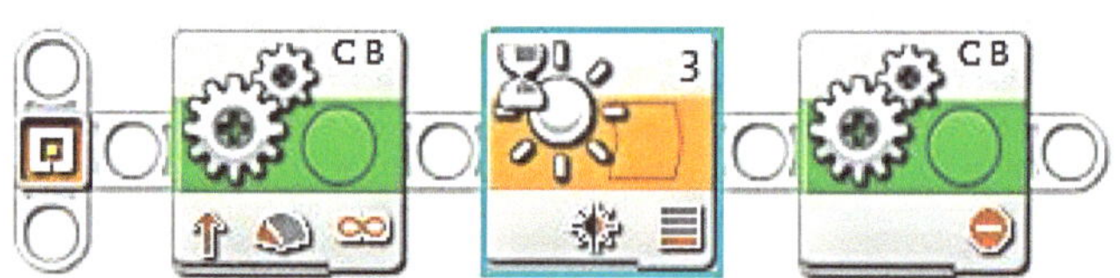

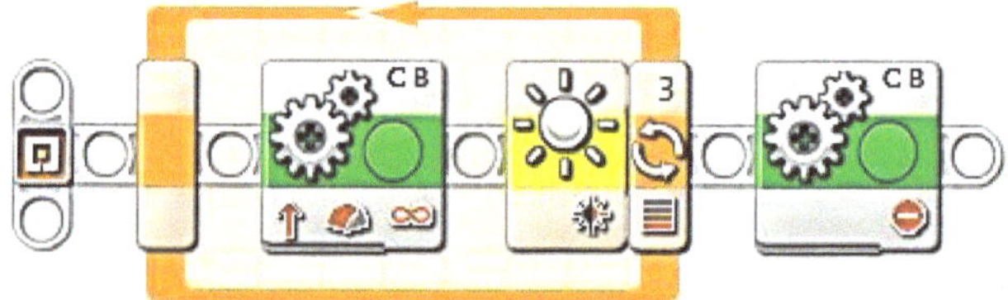

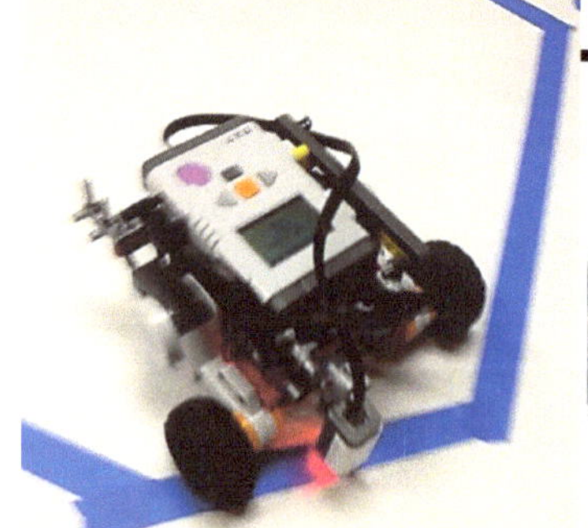

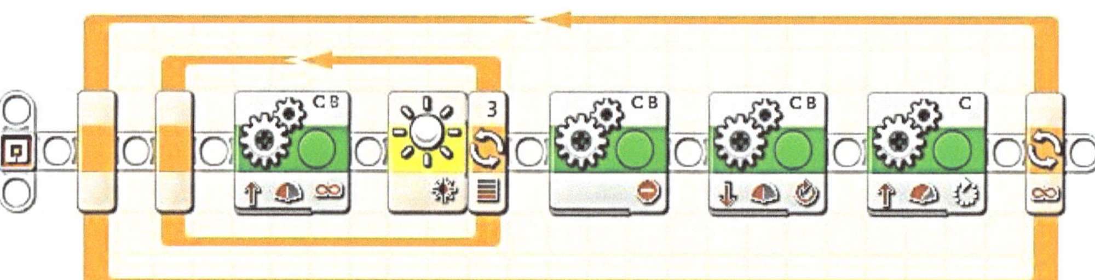

EXERCISE RR Corral:
Corral those robots!

Now you are going to program your robot to go forward forever and do turns. The only thing is that you are going to keep it in a "corral" with fences of just black lines. See if the robot can find its way out of the gate!

Robot Ranch Corral

Gate

Switch Blocks

•A Switch block is a programming block that lets the robot make a decision based on a specific condition that you program - all by itself! When the robot reaches that condition in the running of the program, it will then choose another specific part of the program to then run. This decision making is called a conditional, and the threshold is the value that determines the switch from true or false. You can set the switch block to many different sensors and logic values.

A Switch is like a light switch on the wall - it is either on or off. This can also be known as an "If - Else" statement.

•Below are the Switch blocks set to either the Light sensor or Touch sensor mode. You can see that it opens up 2 white beams for two sets of programs. You then plug in other blocks on these beams, and plug in a condition in the Switch block. You have to make sure that you have only one condition for every switch block.

•<u>LABEL what each part of the program is doing below:</u>

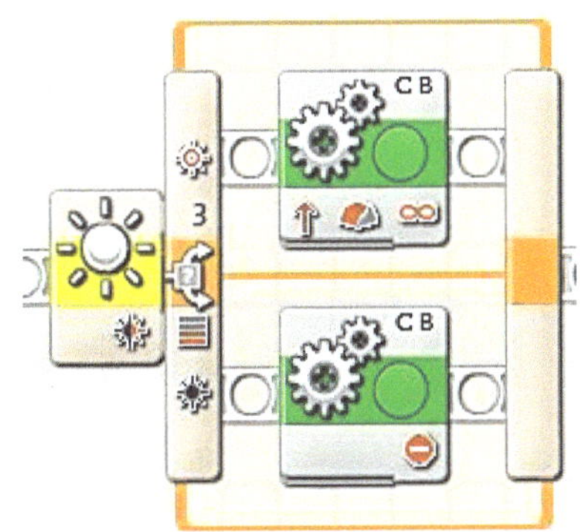

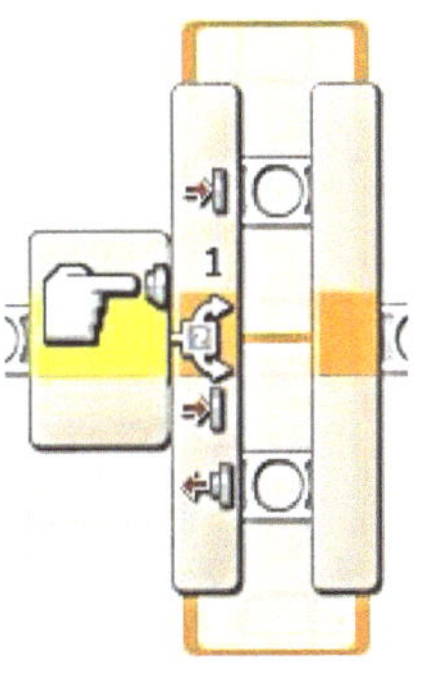

Switch Block Programming Exercises

Switch Exercise: In this mission, you are required to start inside base, and then detect the 3 different wall line check points in a sequential order. You must come back to base. Use a Switch and a LOOP. Tip: set Loop to unlimited, NOT 3x.

Sketch of Manipulator / Sensor: *Draw top view and side view, and label parts:

Draw Path:

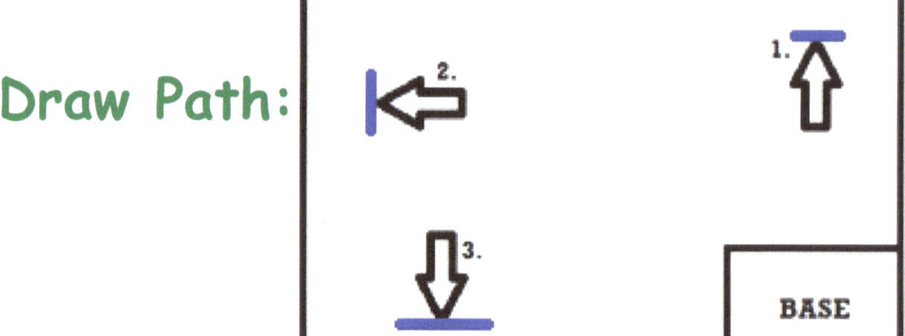

Program File Name: ______________________________

MAKE FLOWCHART:

·Multiple conditions:

·You can place many Switch blocks inside of a Loop block for multiple conditions. Say if you want the robot to stop at a black line OR stop if an object is detected with the touch sensor.

·<u>Multiple Conditions EXERCISE:</u>
Write the following program and run at a box with a black line located 3 inches from the box. Run program. What happens? Then move the line 10 inches from the box. Run program again. What happens? Why is it stopping each time?

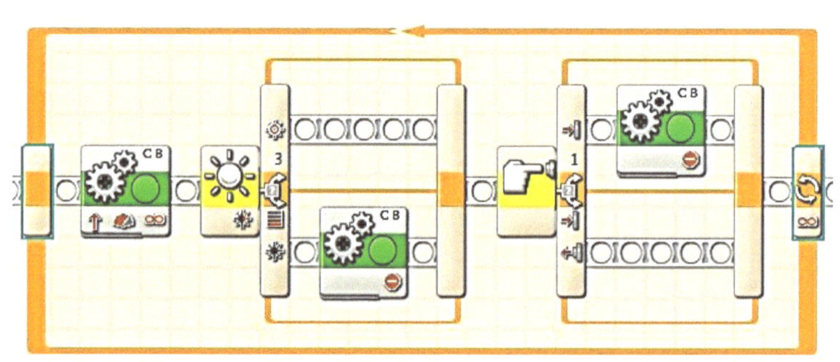

Following a Line with Light Sensors

•By using a line follower, you can just have the robot start and follow along the black or colored lines on the mission field, or along the edge of a carpet, etc. If programmed correctly, they can be so useful to get to difficult places on the mission field.

•We already know how to detect a dark line. We are going to build upon this Light Sensor knowledge to now follow the black or blue line.

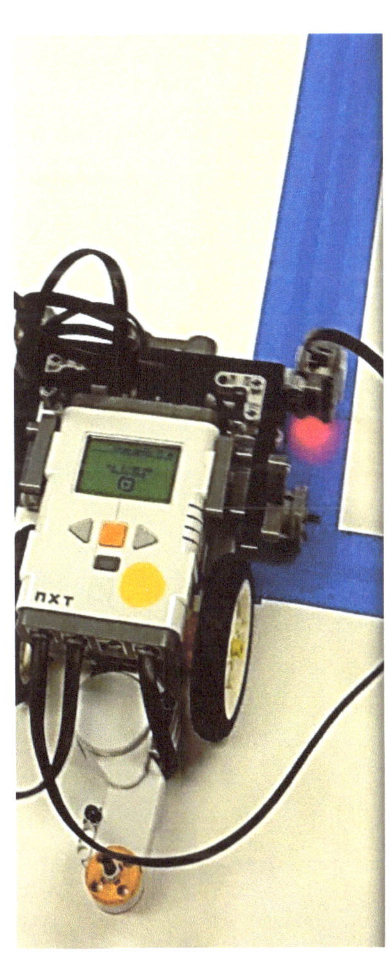

•The most challenging part about the line follower is the programming. It should look something like the program below. It isn't really that hard, you just need to do one step at a time. PAY ATTENTION TO DETAIL! Don't be frustrated if it does not work right away. It usually requires several adjustments!

Program #1: Adjusting steering using both motors at the same time.

•It is a Switch block (light sensor mode) that has Move blocks in it. It is all inside of a Loop block to keep repeating the program.

Line Following Exercises

Program #2: You can also use this line following program. Can you make it go backwards along the line?

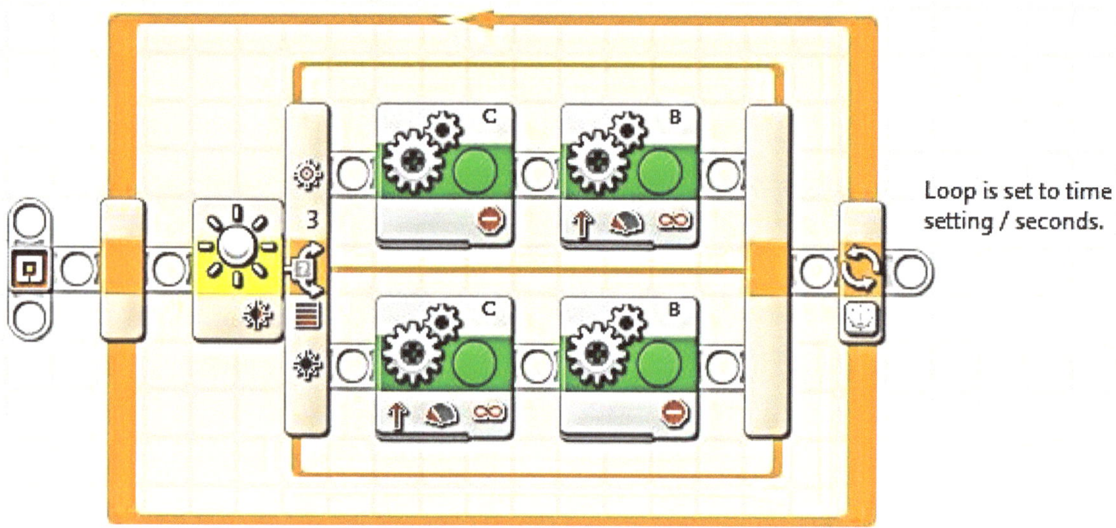

Program #3: This uses Motor blocks. Test out the line following program on our mission field.

This is a line following program following the right side of a line. Please note the power settings for each of the Motor blocks.

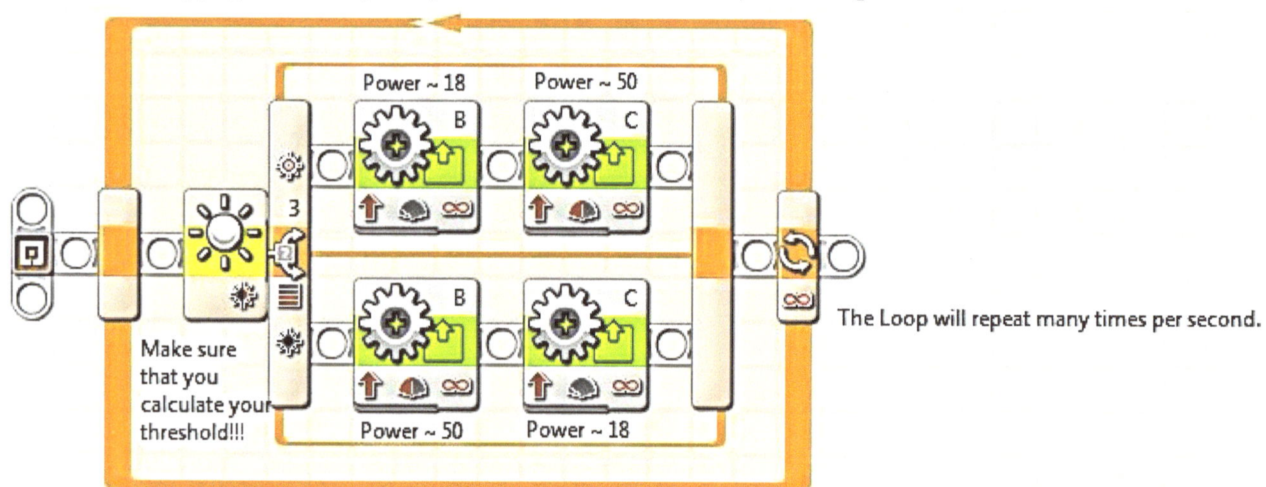

•Which program works best out of the three?

•You can also do line following on inverted line colors!!

Sound Sensors

Sound Sensors act pretty much like a microphone on your robot, and let your robot react to sounds at different volume levels.

A sound sensor is what lets your robot hear, but it is in not anywhere as wonderfully designed and sophisticated like your human ears.

Sound travels as pressure waves, and the pressure waves hit our ear drum, which then transfers through the middle and inner ear until it transfers into nerve impulses which is interpreted by our brain. We can hear all sorts of pitches and differences in sounds, but our robot's sound sensor only hears the difference in volume.

The sound sensor will register sounds on a volume level of 0 - 100. When reading sound measurements on the View, you go to the dB option.

Sound Sensors Measurement Exercise

Try to test out different sounds and sound levels and watch the readings on the screen. Keep the noise source the same distance away from the robot when testing and running missions.

Noise & Level	Distance from Sensor	Measurement reading dB

Using Sound Sensors

To use the sound sensors in your programs, you can use them through the loop, wait, and switch blocks set to the sensor > sound sensor settings.

For the sound sensor to react to a certain sound, you make a measurement reading of the noise you want to trigger the robot in the View. This numerical value is called a THRESHOLD. The threshold is just a value, usually just a number that is the decision point for that command. You usually plug in the value, and then make the conditions either > or < that number on the configuration panel for that block.

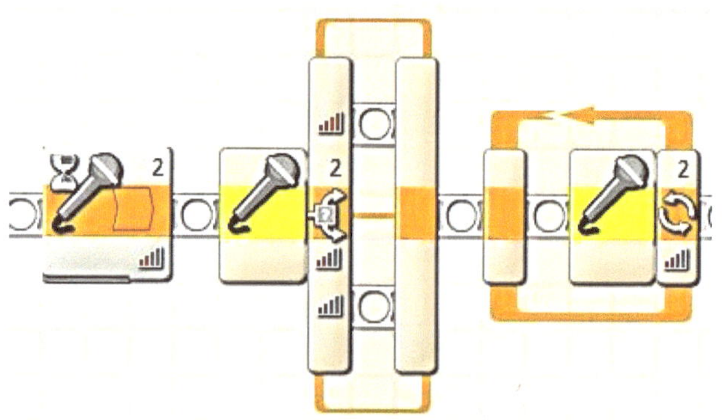

Robo Listen and Follow Exercise

Can you get the robot to listen to you? If the robot hears a soft sound, have it move forward in a straight line. If it hears a loud sound, such as a clap, have it stop and make spins. Can you get it to follow you?

<u>Draw the FLOWCHART of the decision and actions needed for this program below:</u>

START

The Science of Sound

Since we are going to be working with sound blocks and sound sensors, we need to learn and review a little about sound.

What is sound?

Sound is a mechanical wave that is an oscillation (a backward and forward motion) of pressure transmitted through a solid, liquid, or gas.

A wave is a disturbance through space or matter. As the wave travels, energy is transferred from one place to another.

Sound waves are a type of mechanical wave, which travels through a material called the medium. Solids, liquids, and gases are all examples of different media that sound waves travel through. Can a sound travel through a vacuum, such as space?

Types of Waves:

1) Transverse Waves

Motion of medium

wave travels

2) Longitudinal Waves

Compression

(Spring moves back and forth)

rarefraction

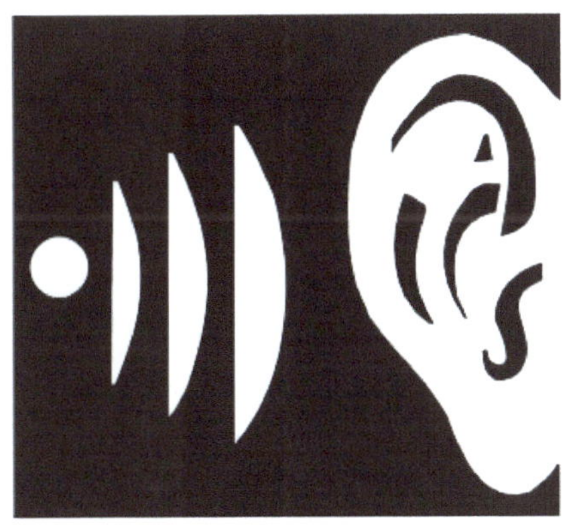

Ripples in water are an example of transverse waves, while sound waves are examples of longitudinal waves.

Volume Measurements

The volume, or loudness, of a sound depends on the amplitude of the sound wave. The more energy you use to make a sound, the greater the amplitude, and thus the louder the sound.

Sound is measured in units called decibels, or dB.

Here are the parts of a wave:

Crest
wavelength
1 Wave
trough
Amplitude

- A wave is made up of 1 crest and 1 trough.
- Frequency – the # of waves in each second

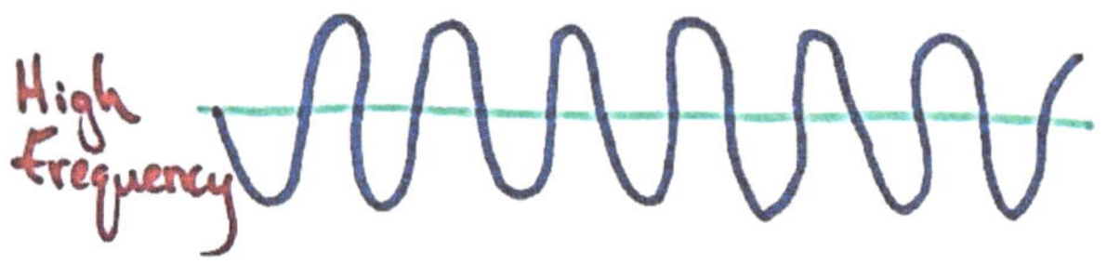

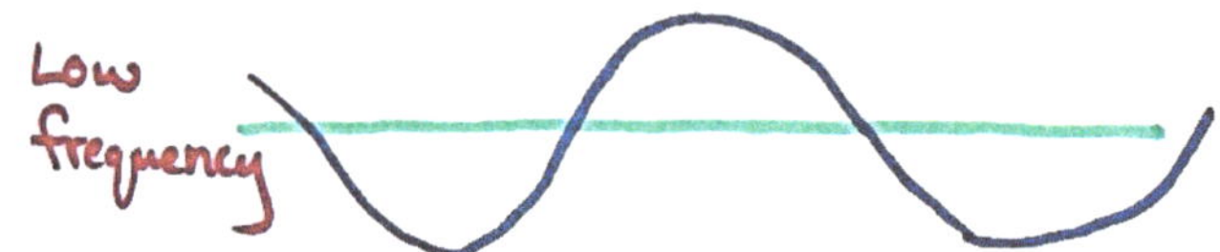

dB	Effect	Volumes of Common Sources of Sound
220	Deafening	NASA's Saturn V rocket at liftoff
180	Deafening	Death of hearing tissue
140	Deafening	Shotgun fire
130	Deafening	(Threshold of terrible pain, decibels at or above 130 cause immediate ear damage) pneumatic rock drill, Jet engine at 100'
120	Deafening	Thunder, fireworks display
110	Deafening	Near-passing train, rock band
100	Very Loud	Passing truck, lawn mower, laying on car horn, circular wood saw
90	Very Loud	Decibels at or above 90 regularly cause ear damage. Motorcycle revving
80	Loud	Noisy office, vacuum cleaner, alarm clock, that noisy kid in class
70	Loud	Average radio, normal street noise
60	Moderate	Conversational speech
50	Moderate	Normal office noise, moderate rainfall
40	Faint	Normal home residence, normal private office
30	Faint	Quiet conversation (this is what your parents want)
20	Very Faint	Whispering, ticking of watch
15	Very Faint	Rustling of leaves in the forest next to a cool relaxing stream with birds.
10	Very Faint	Threshold level of very good hearing

Ultrasonic Sensors

•Ultrasonic sensors are often used in robots for obstacle avoidance, navigation, and map building and are used to detect the presence of targets and to measure the distance to targets in many automated factories and plants.

Ultrasonic sensors (transceivers) -The ultrasonic sensor works by sending out ultrasonic sound waves that bounce off an object ahead of it and then the wave reflects back. It measures the time it took for the echo to determine the distance to an object. The best surfaces that reflect sound waves are hard, flat surfaces. Rounded surfaces reflect the sound waves at different angles away from the dome/ sphere, and soft surfaces can absorbs a lot of the pressure waves.

Ultrasonic—the sound is above the frequencies of audible sound, anything over 20,000 Hz. Audible (normal sounds we hear) are from 20Hz up to 20,000Hz. They are very high frequency waves.

How an ultrasonic sensor works:

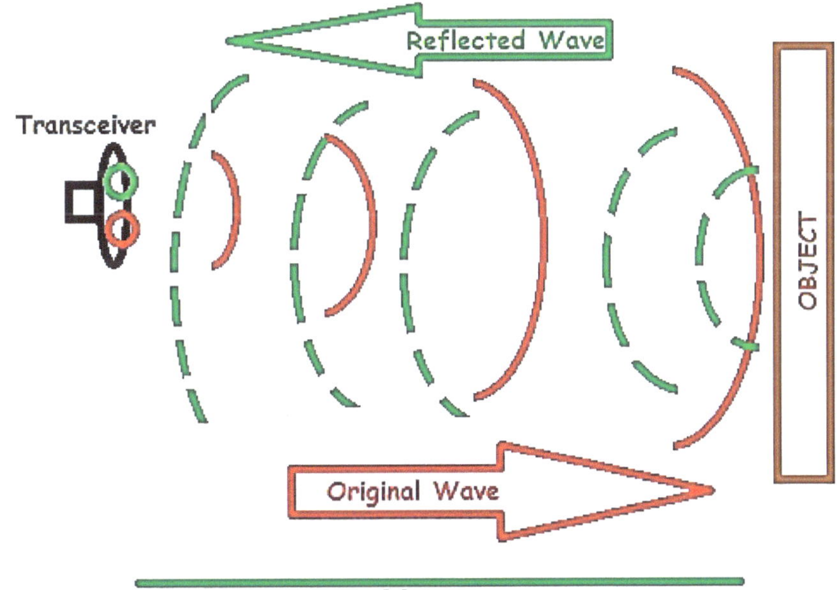

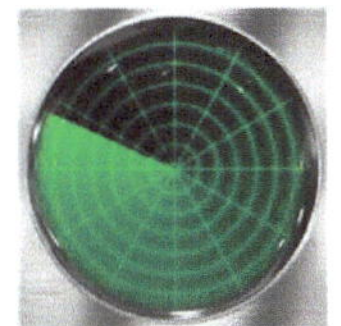

A common use of ultrasound is in range finding, this use is also called SONAR, (sound navigation and ranging). This works similarly to RADAR (radio detection and ranging).

Introductory Ultrasonic Sensor Exercises

Ultra Stop EXERCISE 1:After attaching the ultra sonic sensor via port 4, write a program that runs the robot continuously and stops when the robot is 8 inches from an object or wall. Test it out on the side of a box and measure with a ruler. Does it stop at 8 inches from the object? ____________

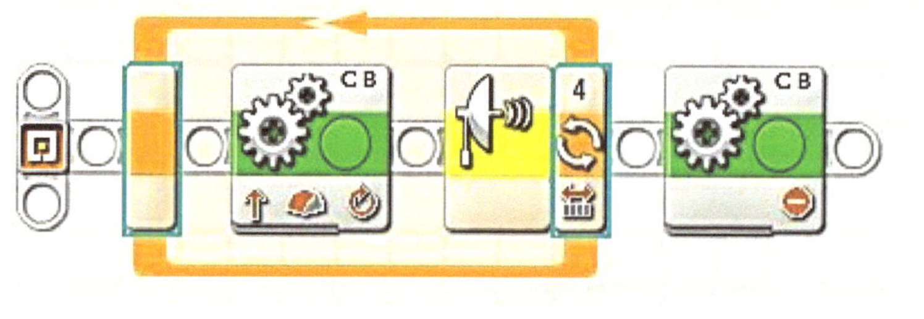

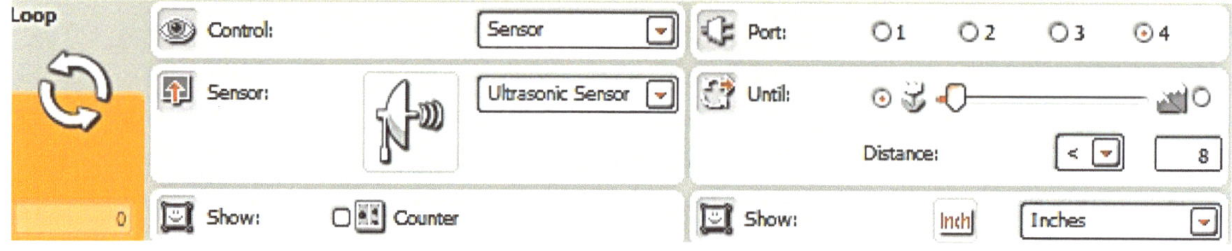

Now try it without the last Move block. How is this different?____________

What other blocks could you use to program this exercise? Do this another way:____________

•Test out the ultrasonic sensor at different angles to the object or wall. Does this affect the stopping distance? Why?____________

•EXERCISE 2: Bumper cars! Now we can do "Non-bumping Bumper cars"!

•Write a program that runs continuously forward until it measures 10 inches from an object. Then have it back and turn, and go forward again. Make sure that the robots do not touch each other, especially when backing up!!

MAKE FLOWCHART:

Parallel to Following Along the Wall Mission

m

ℓ

Parallel lines (L & M) are lines that run along each other but never cross.

In this mission, you are going to have your robot follow along a wall using the ultrasonic sensor. First figure out Path, Draw FLOWCHART, write program, and then run mission!

Program File Name:

______________________.rbt

EXERCISE A: Attach the Ultrasonic sensor to the right side of your robot and pointing to the SIDE and down low. We are then going to have it stay a certain distance from the wall using the ultrasonic sensor, and making movements towards or away from the wall based upon the ultrasonic sensor readings.

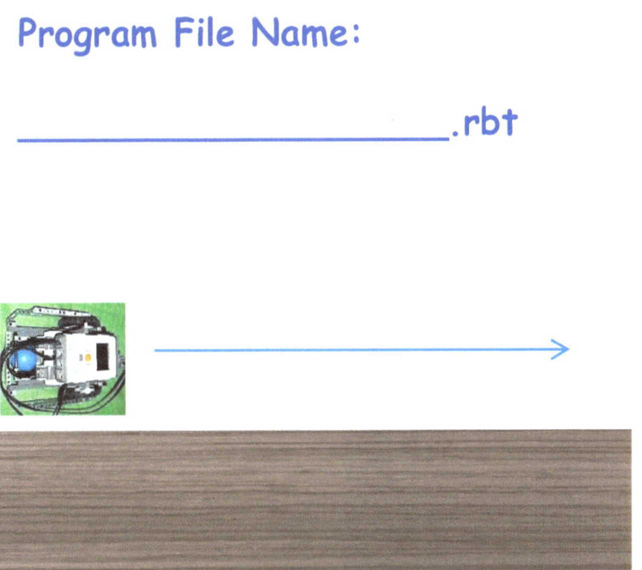

FLOWCHART:

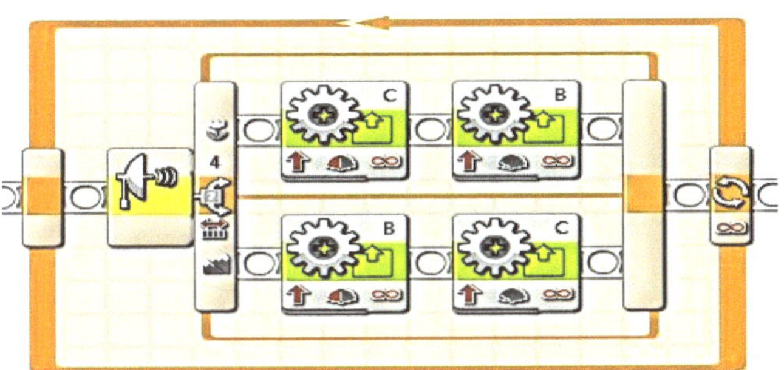

Right Around the Box Mission

EXERCISE B: You are going to have your robot follow along a box using the ultrasonic sensor. You may alter / use the parallel to wall mission.

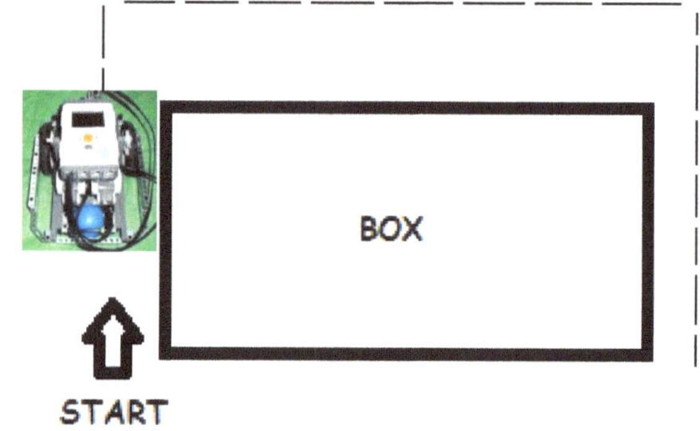

EXERCISE C: Turning a corner! Use a touch sensor to detect that there is a wall in front. First figure out Path, Draw FLOWCHART, write program, and then run mission!

Program File Name:

________________________.rbt

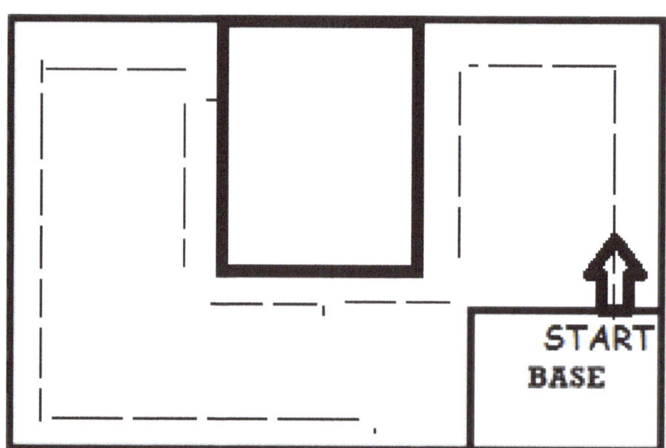

FLOWCHART:

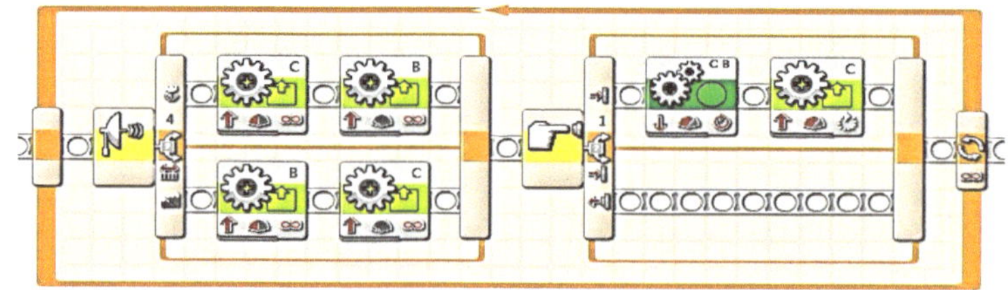

Building a Great & Versatile Robot: Tread Bot!

In our robotics courses, we have built castor bots, swivel bots, skid bots, and box bots, but now we are going into new territory: using TREADS!

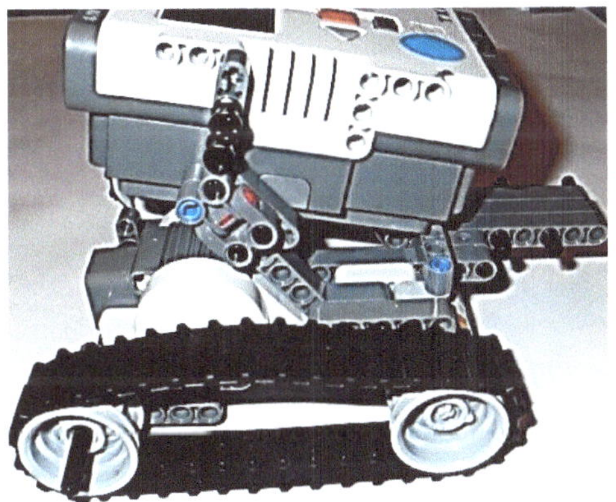

•Your building instructions for this robot are on your computer under:

- ➢Documents
- ➢LEGO® MechaniKids Classes
- ➢Tread Bot Building Instructions

The treads are great for superb traction, stability, and climbing over stuff!

•Treads are going to work differently than our large or small wheels. You will have to go to View to measure out exactly how many degrees the motors need to rotate for the robot to make turns. Also, spins work well when using treads. Have fun! ☺

UNIT 7
NANOTECHNOLOGY ROBOT CHALLENGE

Now in robotics we are going to start on a really, really, really small adventure – the Nanotechology Robot Challenge!

This Nanotech Challenge is a robotic mission challenge based around a unit study about the field of nanotechnology, You will be completing robotics missions that all are based on a nanotechnology theme, and will learn about this exciting scientific field.

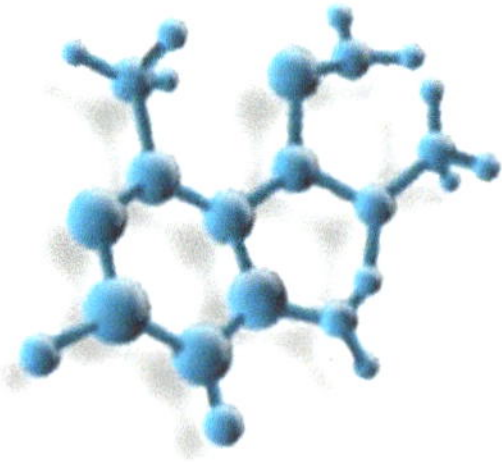

Introduction to Nanotech Challenge

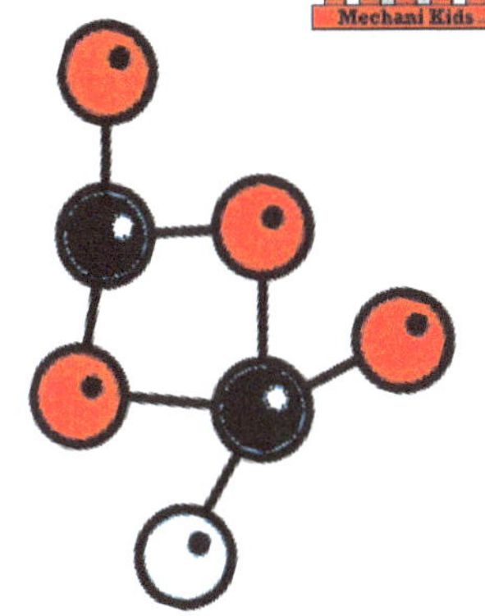

What is Nanotechnology?

Also called nanotech, nanotechnology is the study of manipulating matter on an atomic and molecular sized level. The field of nanotechnology deals with structures sized between 1 to 100 nanometers, and involves developing devices or materials at that size.

A *nanometer* is one billionth of a meter, around the size of three atoms together.

Compare this with the size of the average human hair: about 25,000 nanometers wide.

•Nanotechnology may be able to invent useful new materials and devices for the fields of electronics, medicine, and energy production. However, nanotechnology raises many of the same concerns as any other new technology, including fears about the environmental impact of nanomaterials.

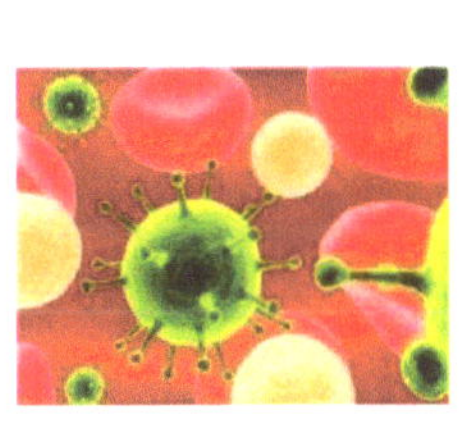

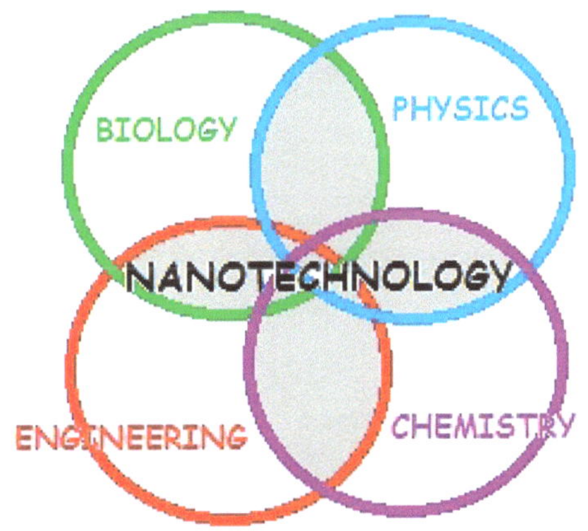

•The field of nanotechnology is a compilation of the different science fields of biology, physics, chemistry, and engineering.

•One nanometer (nm) is one billionth, or 10^{-9}, of a meter.

•A DNA double-helix has a diameter around 2 nm.

NANOTECH MISSION MAP

Here are the elements that we are going to build for the different task missions for the Nano Tech Challenge:

1) Nano Stain Resistant Fabric
2) ATP Molecular Motor
3) Atom Self Assembly
4) Individual Atom Manipulation Shaker Table
5) Unstick NanoTip
6) Bone Medicine
7) CNT Truck Lift
8) Pizza Scent Molecules
9) Space Elevator

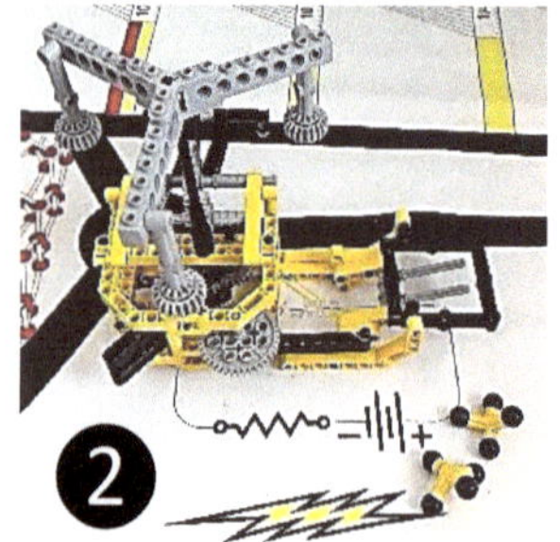

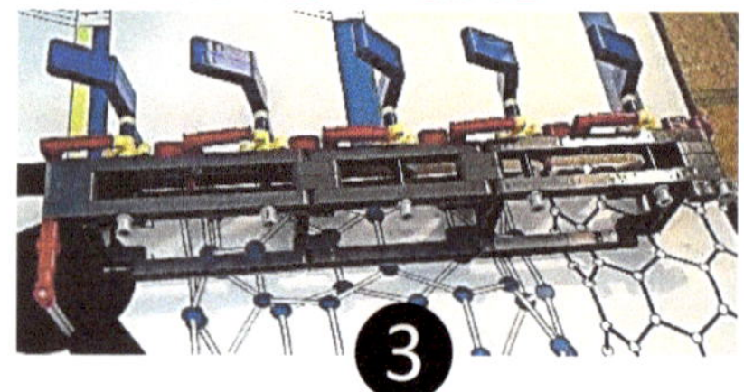

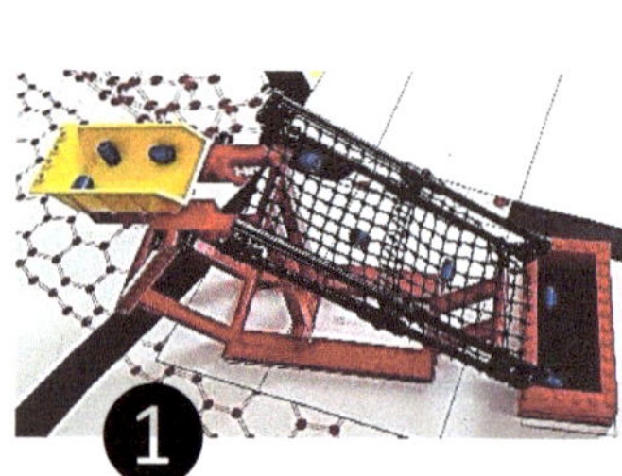

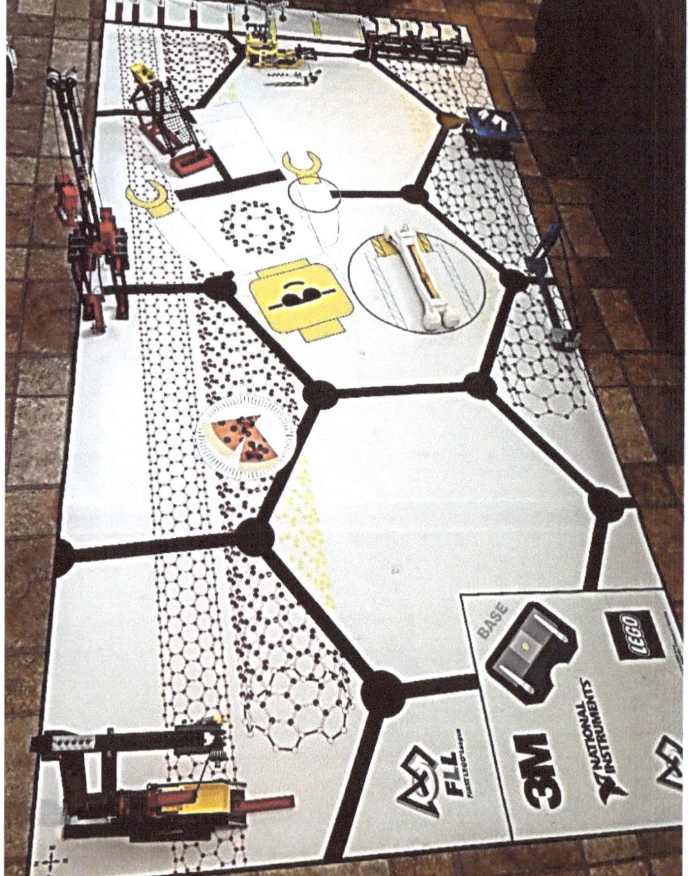

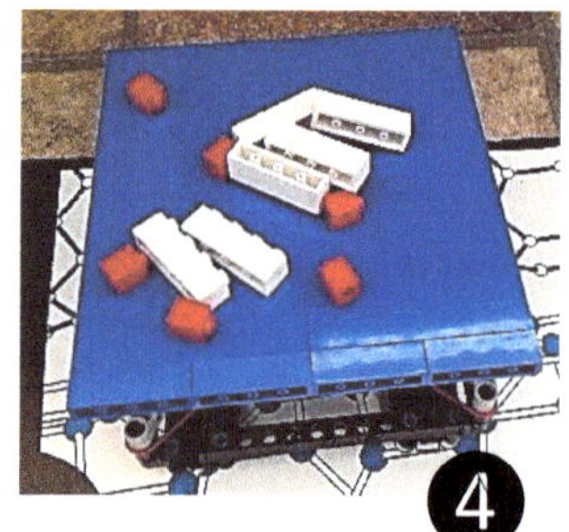

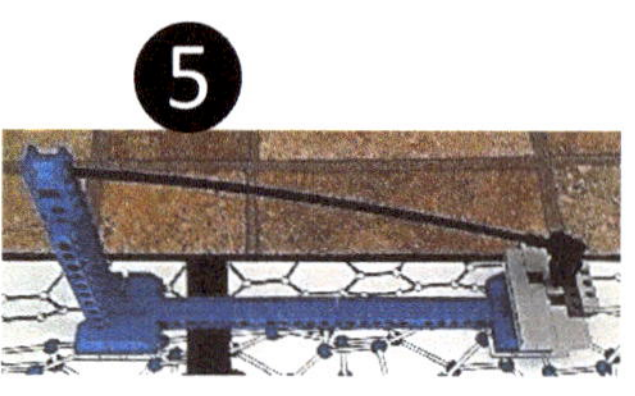

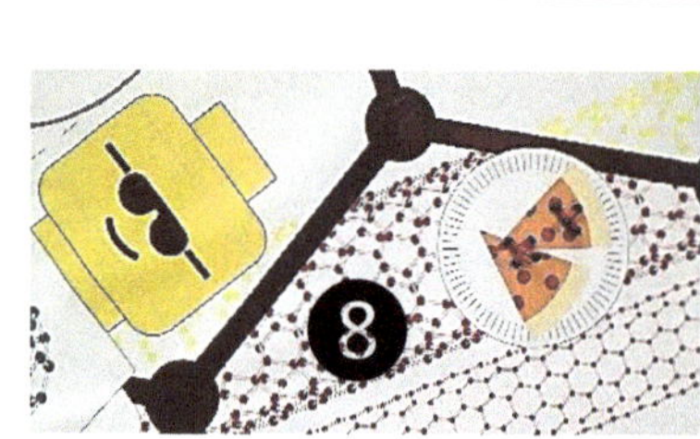

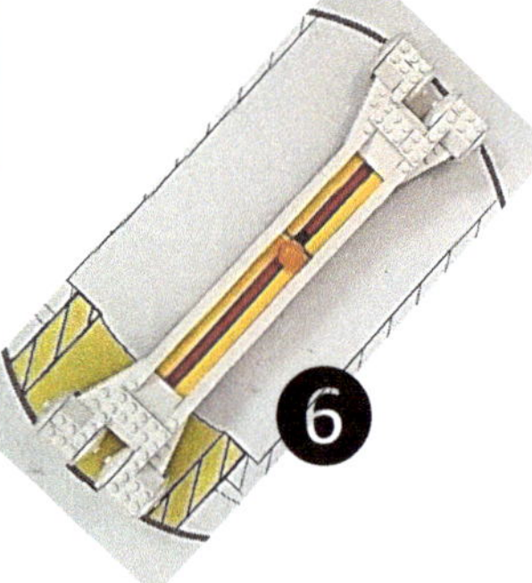

Mission Table Overview

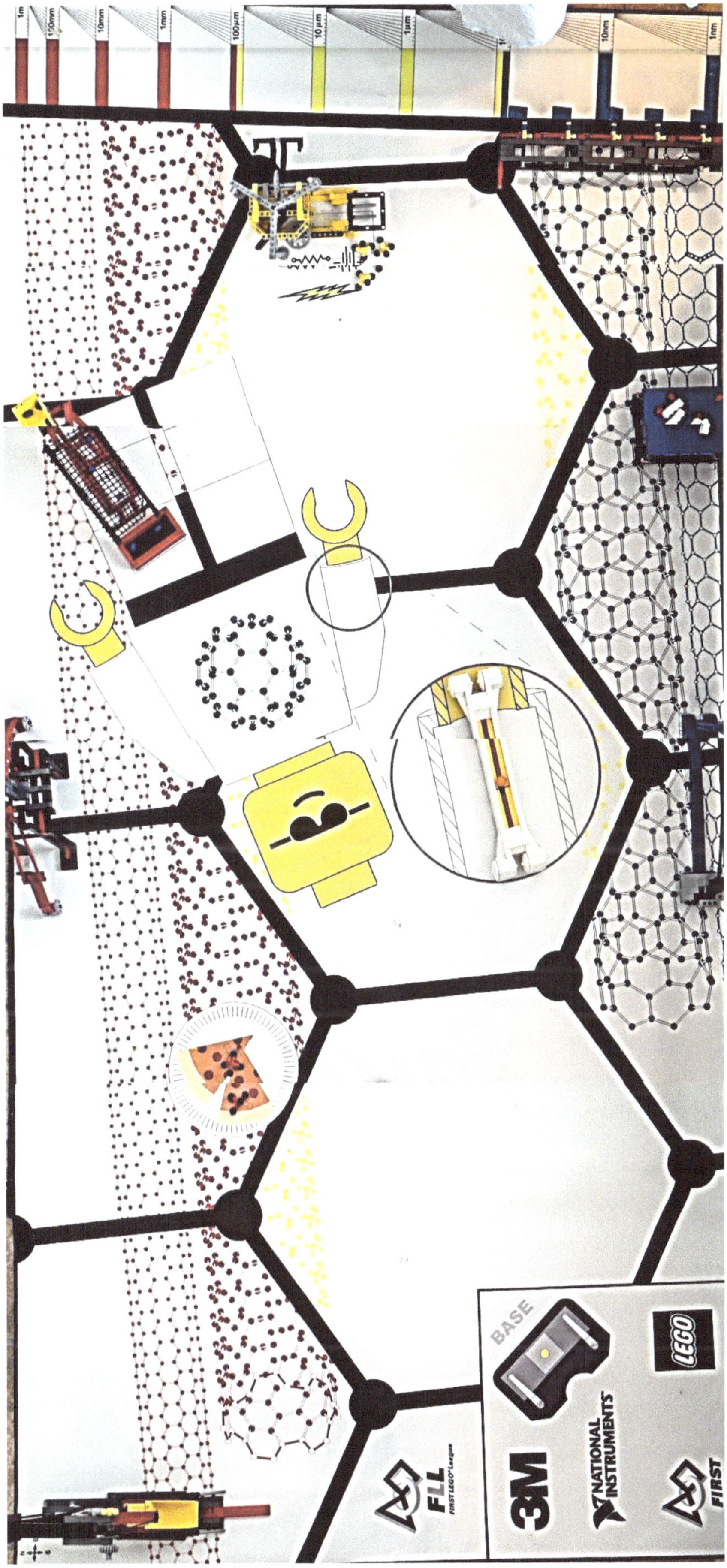

How to Complete the Nanotech Mission Challenge Worksheets

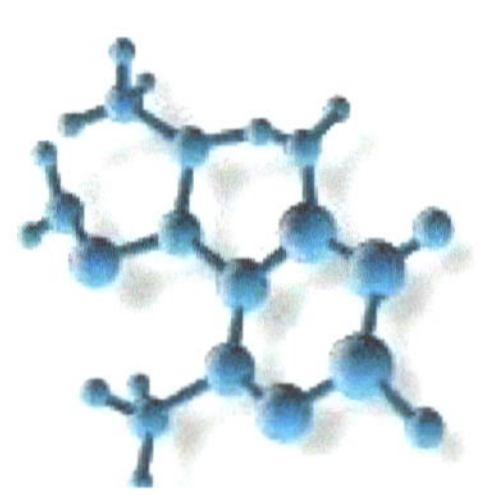

Here is how you need to fill out mission assignment sheets to have your points from your mission runs count. Follow the order stated below:

ALL PARTS OF THE MISSION ASSIGNMENT SHEETS MUST BE COMPLETELY FILLED OUT FOR YOU TO MOVE ON TO THE NEXT MISSION! *Also, please note that it may take more than one class session to complete these missions.

1) When you start your mission, you need to write down your name, your teammate's name, the Class #, and your PC's/robot's name in the red box at the top right corner of the page. Leave the rest blank until later.

2) Read over the mission information, and the background of the mission.

Mission Completion Information

Instructor Check:

Name:

Teammate :

Robot/Computer:

Class (es):

File Program name:

Total Mission Points:

Nanotech Robotics

Page 22

Bone Medicine

Mission: Put the medicine in a specific location. The Buckyball containing medicine must be released into the person's arm bone. 50 points are awarded if you place the Buckyball in the yellow/red bone marrow and it rolls onto the cancer spot.

Background: Cancer happens when damaged cells in the body start growing and spreading fast. These cancer cells are similar to healthy cells, which makes cancer treatment difficult for doctors. Treating cancer tends to kill some of the cells which are healthy along with the cancer cells, which results in patients becoming sick. What is needed by patients is treatment that is designed to target only cancer cells and leave the healthy cells unharmed. The solution to this problem may come from Nanotechnology. When we are given medicine, the medicine circulates through the body and causes harmful side effects in unplanned areas. Nanotechnology may potentially allow some medicines to be precisely placed inside special molecules like the C60 Buckyball molecules to enable transfer of the medicine to the precise area where the medicine is needed, instead of harming the entire body.

Location of Target & Mission Path:

Draw Manipulator:

Program Flowchart:

Sensors Used:

Motors used:

3) Next, DRAW your target locations on map. Complete the 3 P's of robotic mission solving: Path, Program, and Pieces. Make sure to DRAW your path on the mission map area!

4) Write your computer program. WRITE DOWN the program's file name up in the red box at the top right corner of the page.

5) When mission is completed, DRAW a sketch of the manipulator on the graph paper area, WRITE down the sensors used and motors that were used.

6) Flowchart your path / program.

7) After you have done all of this, take it to your teacher for him/her to check off your mission and to put the completed points. ONLY YOUR TEACHER CAN DO THIS.

Mission Completion Information

Instructor Check: ☐

Name: ____________________

Teammate :____________________

Robot/Computer:____________

Class (es): ______________

File Program name:

____

Total Mission Points: ☐

Nanotech Robotics

Bone Medicine

Mission: Put the medicine in a specific location. The Buckyball containing medicine must be released into the person's arm bone. 50 points are awarded if you place the Buckyball in the yellow/red bone marrow and it rolls onto the cancer spot.

Background: Cancer happens when damaged cells in the body start growing and spreading fast. These cancer cells are similar to healthy cells, which makes cancer treatment difficult for doctors. Treating cancer tends to kill some of the cells which are healthy along with the cancer cells, which results in patients becoming sick. What is needed by patients is treatment that is designed to target only cancer cells and leave the healthy cells unharmed. The solution to this problem may come from Nanotechnology. When we are given medicine, the medicine circulates through the body and causes harmful side effects in unplanned areas. Nanotechnology may potentially allow some medicines to be precisely placed inside special molecules like the C60 Buckyball molecules to enable transfer of the medicine to the precise area where the medicine is needed, instead of harming the entire body.

Location of Target & Mission Path:

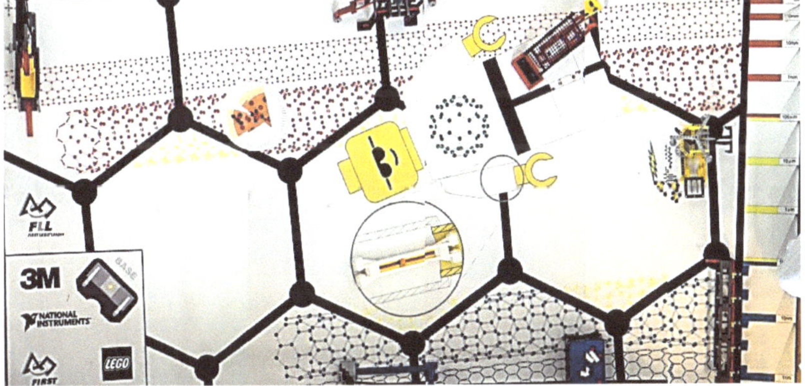

Draw Manipulator:

Program Flowchart:

Sensors Used:____________ Motors used: ______________

Mission Completion Information

Instructor Check:

Name: ____________________

Teammate :________________

Robot/Computer:____________

Class (es): _______________

File Program name:

Total Mission Points:

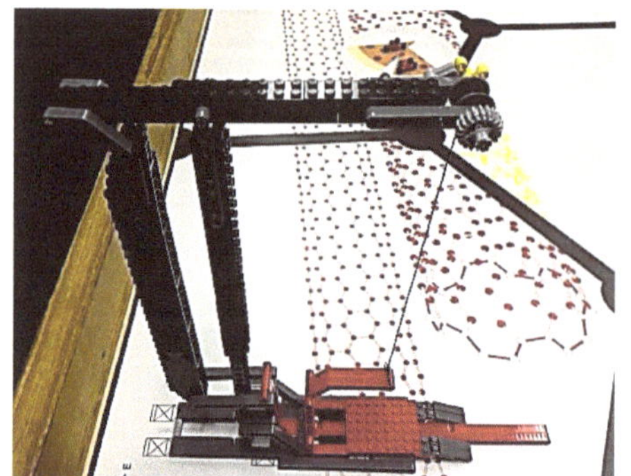

Nanotech Robotics

Carbon Nanotube (CNT) Truck Lift

Mission: Lift the dump truck by a thin cable of carbon nanotubes. The robot must move the dump truck onto the lift frame and trigger the lift frame.

Background: Nanotechnology researchers are studying the carbon atom because carbon atoms can be organized to form carbon nanotubes. These carbon nanotubes can be used as the foundation of some very strong materials. For example, think of a carbon nanotube cable as thin as string which is stronger than steel and can support a ton.

Location of Target & Mission Path:

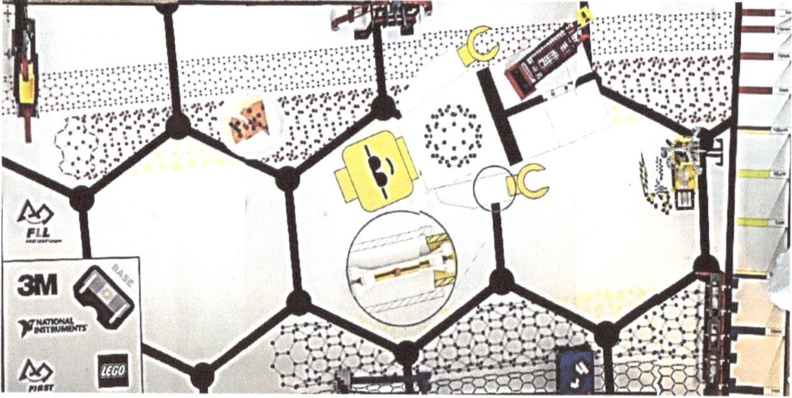

Draw Manipulator:

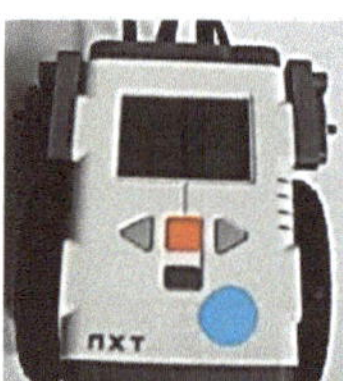

Program Flowchart:

Sensors Used:____________ Motors used: ______________

Mission Completion Information

Instructor Check:

Name: ______________________

Teammate :_________________

Robot/Computer:____________

Class (es): ________________

File Program name:

____

Total Mission Points:

Nanotech Robotics

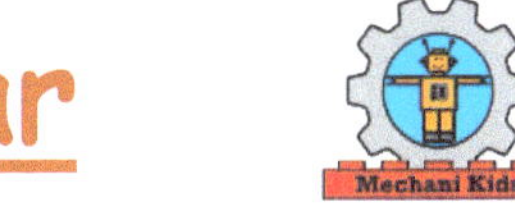

ATP Molecular Motor

Mission: Transport an adenosine triphosphate (ATP) molecule to a molecular motor which will power it and cause it to spin and release energy. You must use the robot to drop 2 ATP molecules into the molecular motor's black square side frame.

Background: Atoms and molecules are hard to work with because they are constantly moving, but if a special molecule is spun a certain way, it can be used to do work. Molecular motors are molecules which have the ability to change other molecules' chemical energy into rotational energy to do work on a small scale such as moving molecules or tightening muscles.

Location of Target & Mission Path:

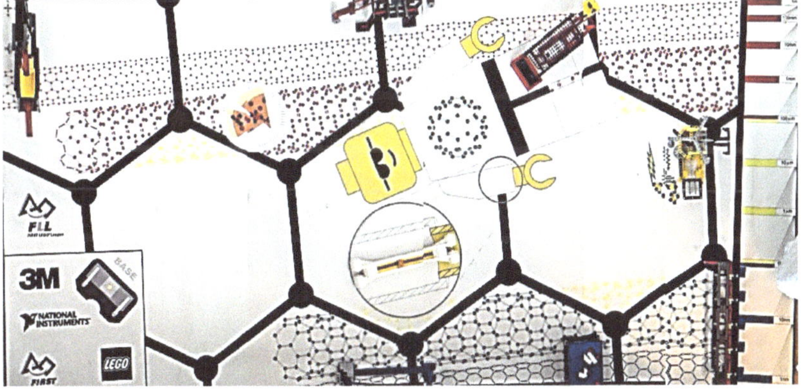

Draw Manipulator:

Program Flowchart:

Sensors Used:____________ Motors used: ______________

Nanotech Robotics

Atom Manipulation

Shaker Table

Mission Completion Information

Instructor Check: ☐

Name: ____________________

Teammate :__________________

Robot/Computer:____________

Class (es): _________________

File Program name:

____

Total Mission Points: ☐

Mission: Move individual atoms accurately. The robot must remove at least 2 white atoms and 2 red atoms from the shaker table surface. Each atom vibrated off table is worth 25 points.

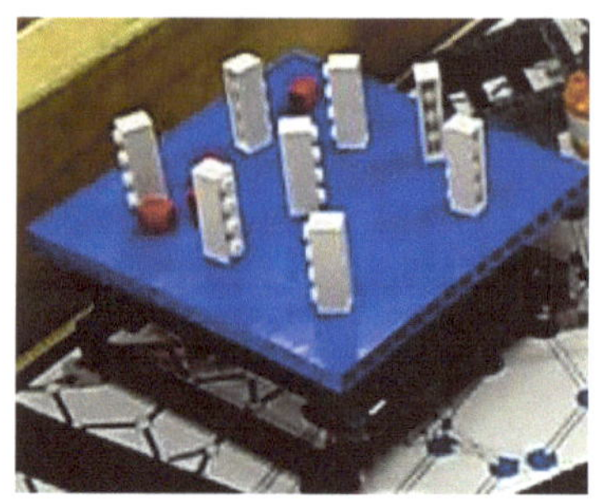

Background: With nanotechnology, materials made up of matter can now be assembled or moved by each individual atom into new uses. Working on the nano size, where materials are about 100, is difficult because everything is vibrating rapidly and shaking.

Location of Target & Mission Path:

Draw Manipulator:

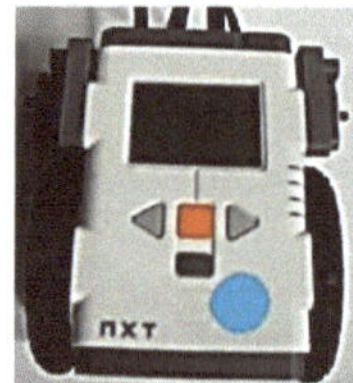

Program Flowchart:

Sensors Used:____________ Motors used: ______________

Nanotech Robotics

Unstick NanoTip

Mission Completion Information

Instructor Check: ☐

Name: ____________________

Teammate : ________________

Robot/Computer: ____________

Class (es): ________________

File Program name:

____

Total Mission Points: ☐

Mission: Unstick the microprobe's nanotip. The robot must separate the nanotip from the material surface. Nanoprobe tip must be completely unstuck from surface, and then reattach it: 40 points.

Background: Atomic Force Microscopy

Just like how we can differentiate different surfaces by feeling if a surface is rough, smooth, bumpy, or sticky, an atomic force microscope can describe a nanosurface atom by atom by the use of its probe or nanotip. The problem with this practice is that these extremely small probes get stuck to the nanosurface a lot of the time because of the strong forces at the nanolevel. This is a problem for many nanoscientists.

Location of Target & Mission Path:

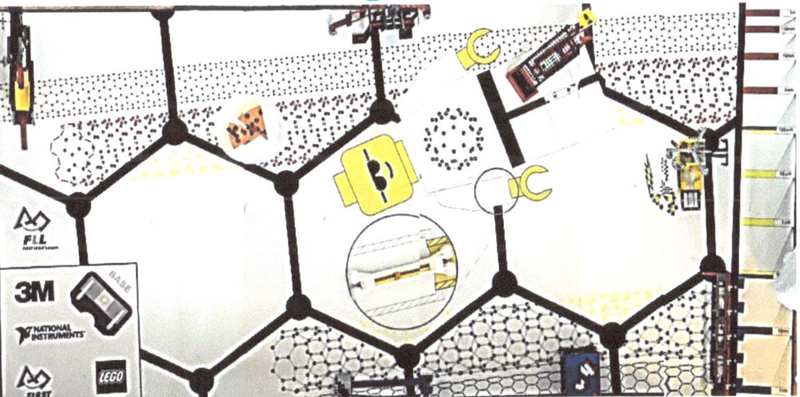

Draw Manipulator:

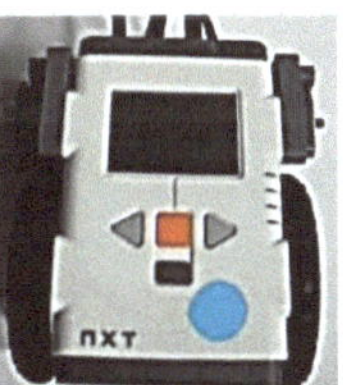

Program Flowchart:

Sensors Used: ____________ Motors used: ____________

Nanotech Robotics

Mission Completion Information

Instructor Check: ☐

Name: ______________________

Teammate :__________________

Robot/Computer:_____________

Class (es): ________________

File Program name:

____

Total Mission Points: ☐

Nano Stain Resistant Fabric

Mission: The robot must deliver from base the dirt trap to the fabric at its location mark on map and then "spill" the blueberry stains on fabric by dumping them on the fabric. Each blue stain molecule is worth 10 points if it falls inside the dirt trap.

Background: Nanotechnology is making a new fabric that has the lotus effect, which imitates the lotus plant leaf where water and stains "roll" right off the fabric due to the fabric's micro-weave properties! The liquid stain forms a water bead on the fabric's area that then rolls off instead of being absorbed into the material.

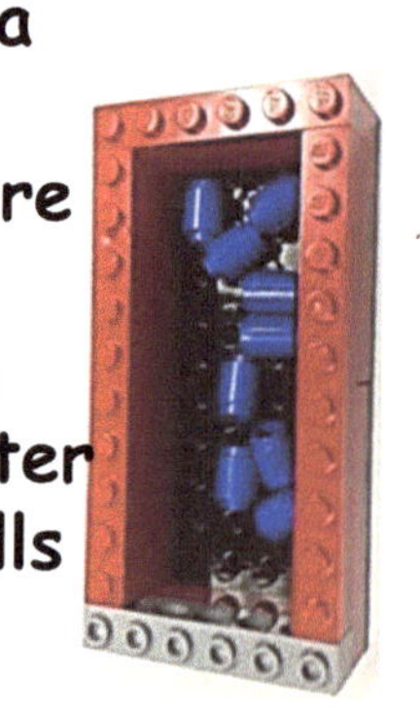

Location of Target & Mission Path:

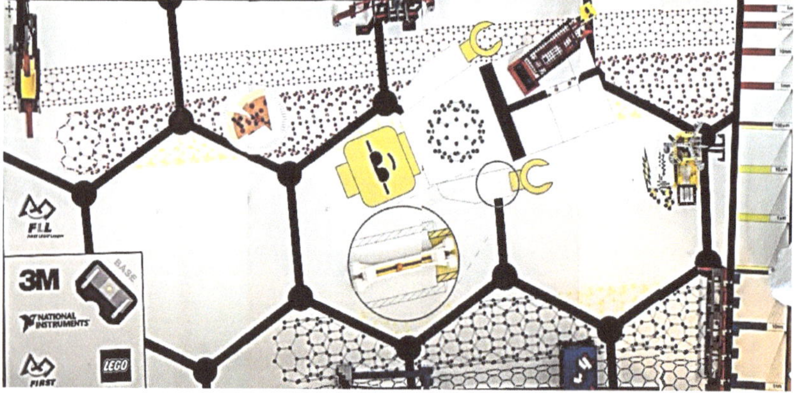

Draw Manipulator:

Program Flowchart:

Sensors Used:____________ **Motors used:** ______________

Mission Completion Information
Instructor Check: ☐
Name: ____________________
Teammate :________________
Robot/Computer:____________
Class (es): ________________
File Program name:
____
Total Mission Points: ☐

Nanotech Robotics

Pizza Molecules

Mission: Gather the scent molecules from the pizza and deliver them towards the minifig's nose. The robot must get the two pizza molecules completely off the paper plate for 10 points each, and transferred to the yellow or black areas of the minifig's face (between the gray lines) for an additional 10 points each.

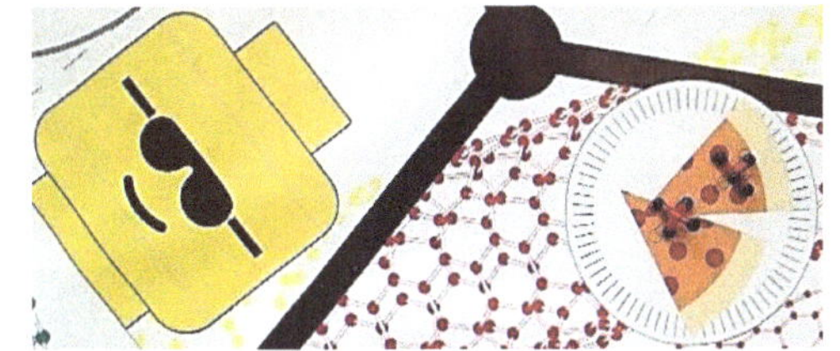

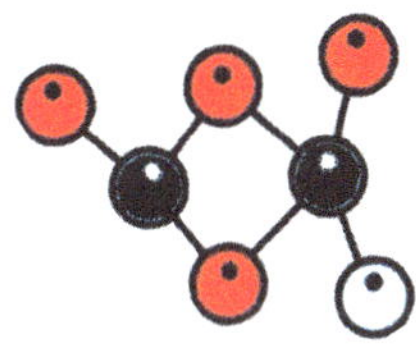

Background: Your nose was designed with the capability to differentiate between lots of different smells. Your nose does this chemically by sensing the different types of molecules from the thing you are smelling. Small molecules that are around 1 nanometer (nm) tend to become gases. The thought of pizza is making me hungry, I'm almost imagining that I can smell those pizza molecules! ☺

Location of Target & Mission Path:

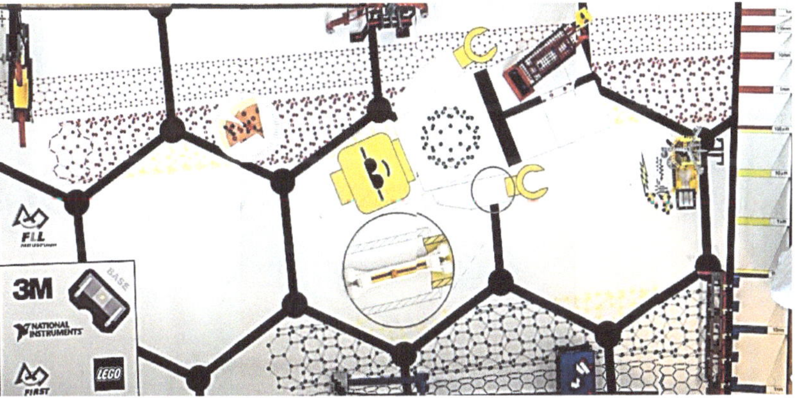

Draw Manipulator:

Program Flowchart:

Sensors Used:____________ Motors used: ______________

Nanotech Robotics

Atom Self-Assembly

Mission Completion Information

Instructor Check: ☐

Name: ____________________

Teammate :__________________

Robot/Computer:______________

Class (es): _________________

File Program name:

____

Total Mission Points: ☐

Mission: Kick start the self-assembly of the atoms so that they line up. The robot must move the lever to cause a chain movement of the blue nanotubes to swing into a long chain. 70 points.

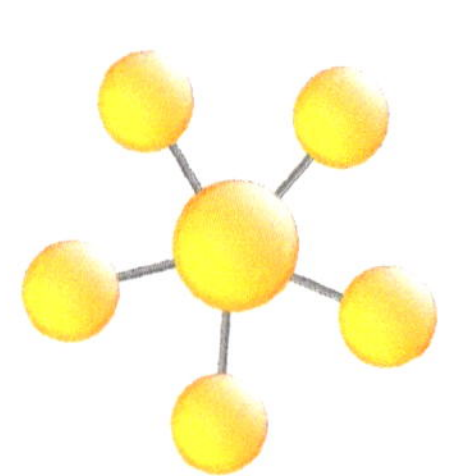

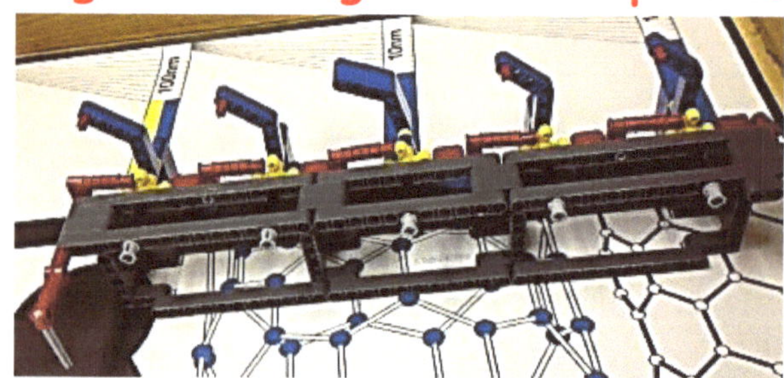

Background: Atoms are tiny, so manipulating them one at a time is extremely difficult. Nanotechnology scientists are trying to develop practices that cause small matter to arrange itself, kind of like magnetic fields do.

Location of Target & Mission Path:

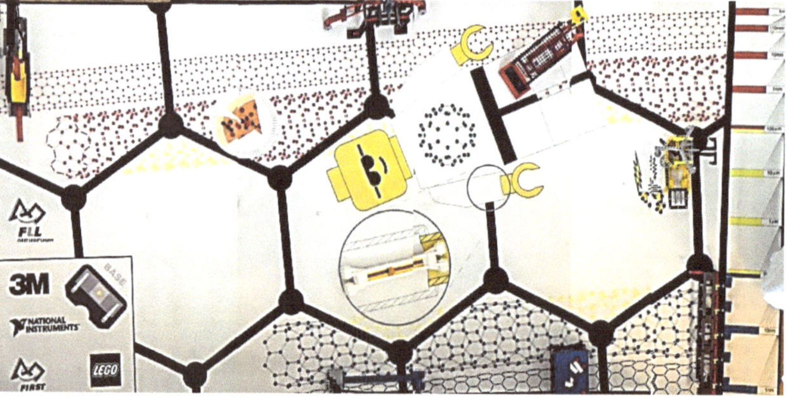

Draw Manipulator:

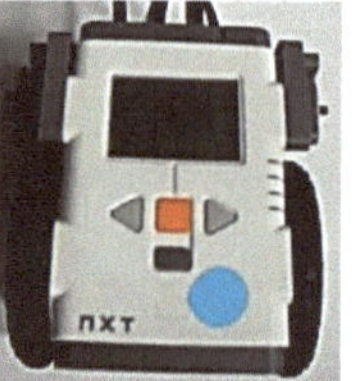

Program Flowchart:

Sensors Used:____________ **Motors used:** ______________

Mission Completion Information

Instructor Check: ☐

Name: ______________________

Teammate :_________________

Robot/Computer:____________

Class (es): ________________

File Program name:

____

Total Mission Points: ☐

Nanotech Robotics

Space Elevator (Carbon Nanotube)

Mission: Activate the space elevator to go to space! The robot must cause the car with the yellow cargo to come down. You must use a sensor. Points: 55.

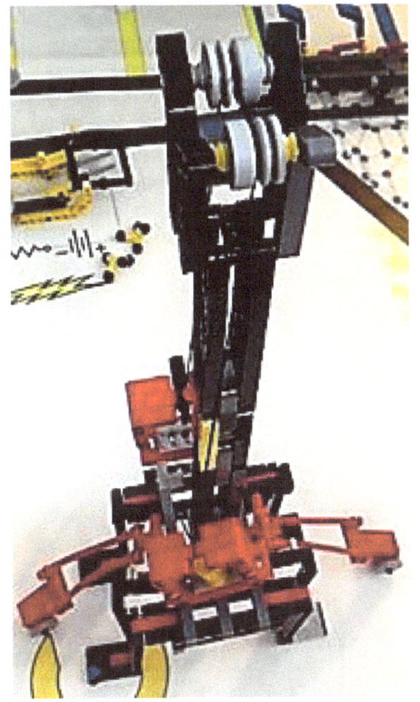

Background: Due to how expensive it is for NASA to send missions to space via the use of rockets, there is an idea that we could create a space elevator by having a counterweight satellite out in space connected to earth with a carbon nanotube cable. Then cargo and astronauts could catch a ride on an elevator to get out of our atmosphere! Although this is just a futuristic idea, some scientists think that it could be done with advances in nanotechnology.

Location of Target & Mission Path:

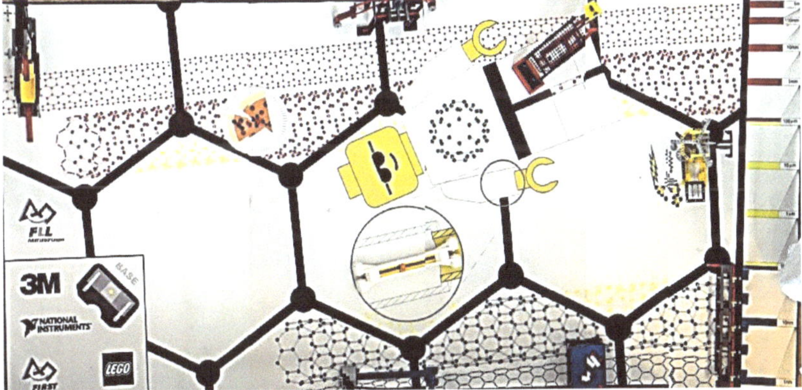

Draw Manipulator:

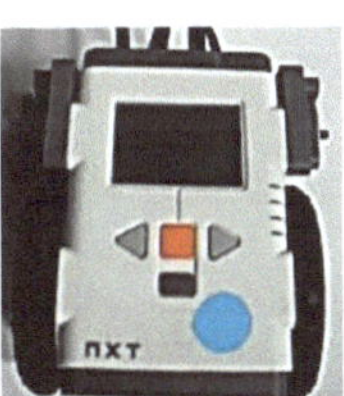

Program Flowchart:

Sensors Used:_______________ **Motors used:** _______________

UNIT 8
ADVANCED PROGRAMMING EXERCISES

In this unit, you are going to do more in-depth programming. The best way to make complicated programming more simple is by starting off with a good flowchart. These programming exercises will teach you how to program the robot so that it is more autonomous, which is the ultimate goal of robotics!

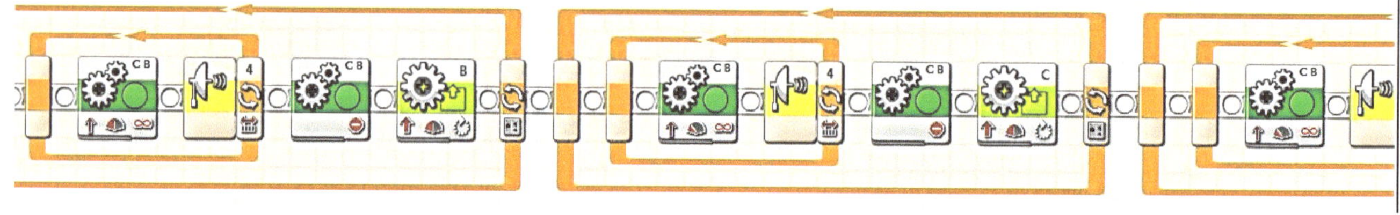

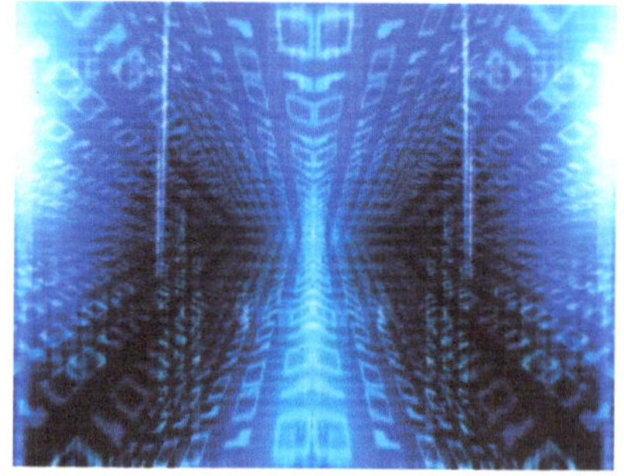

Extended Touch Sensor Exercises

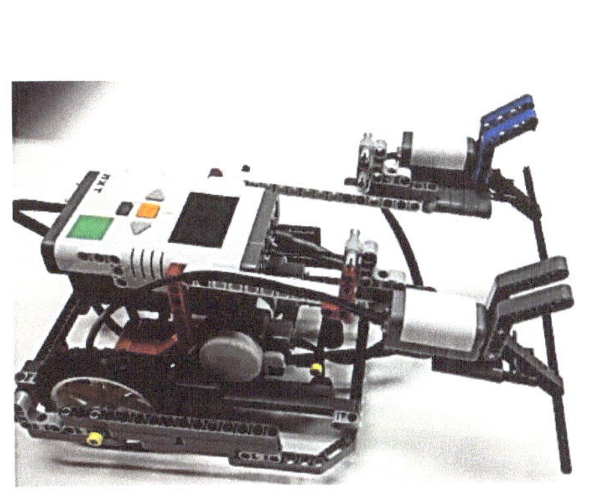

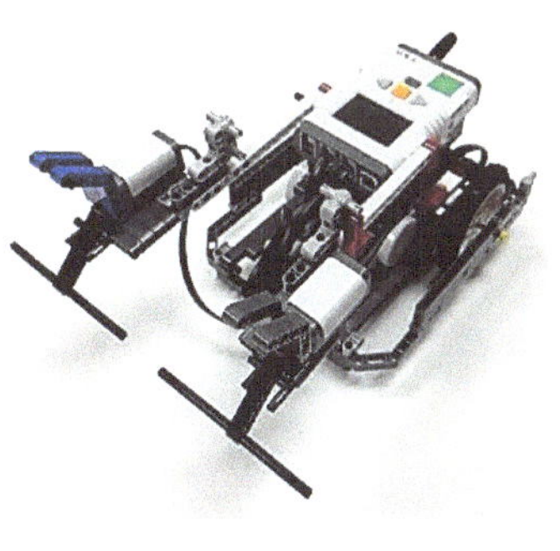

·<u>EXERCISE: Use 2 Bumpers.rbt!</u>

·Use 2 touch sensors to build a right and left bumper. Plug one into Port 1 and the 2nd into Port 2. ***Keep their names in mind when programming!!!! You will need to use switches!

Task Mission: Have the robot go forward and if the L bumper is pressed, have the bot stop, back up, and turn to the right. If the R bumper is touched, have the robot stop, back up, and make a left turn. Once you get it to run well on the table, run it down a hallway to see if it can navigate the hallway and get around corners! Here is a flowchart of the actions and program:

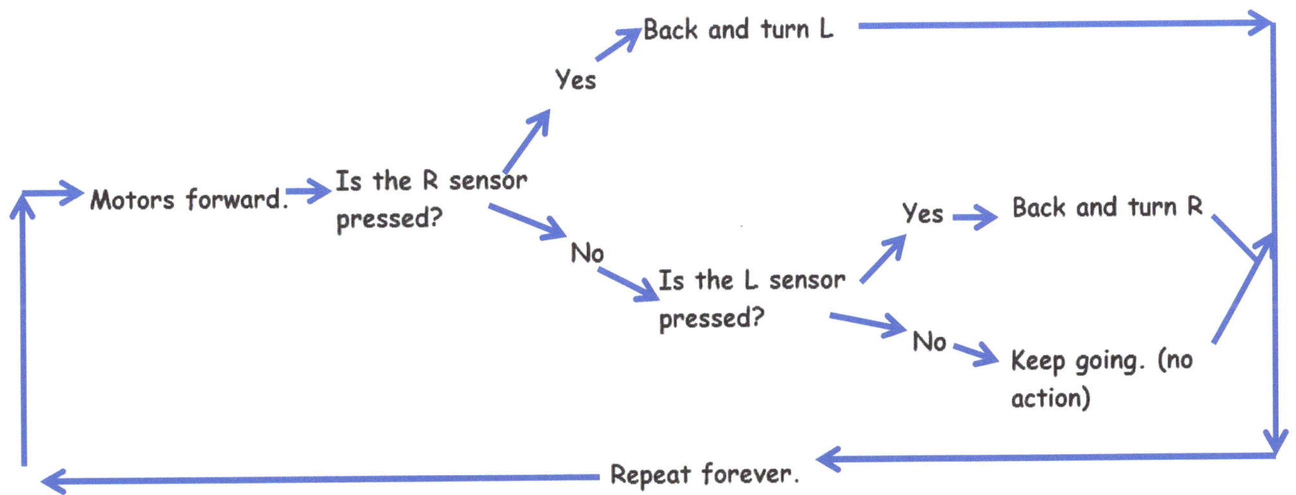

Pseudocode

Flowcharting is a great way to plan out the robotic missions. Once you have finished flowcharting and make sure that it makes sense, you get ready to write your program. Once you write the program, you can comment out code, which explains each part of the program on the programming workspace.

Then, after you write your program, you can do something called pseudo-coding, which is written out text that describes the program in a simple, vertical, orderly fashion. It makes a description of the program that is very detailed, yet logical and short.

Pseudocode looks like a real text-based program but does not follow any strict rules, especially punctuation rules. Here is an example for a mission that has the robot go forward and if the L bumper is pressed, have the bot stop, back up, and turn to the right. If the R bumper is touched, have the robot stop, back up, and make a left turn.

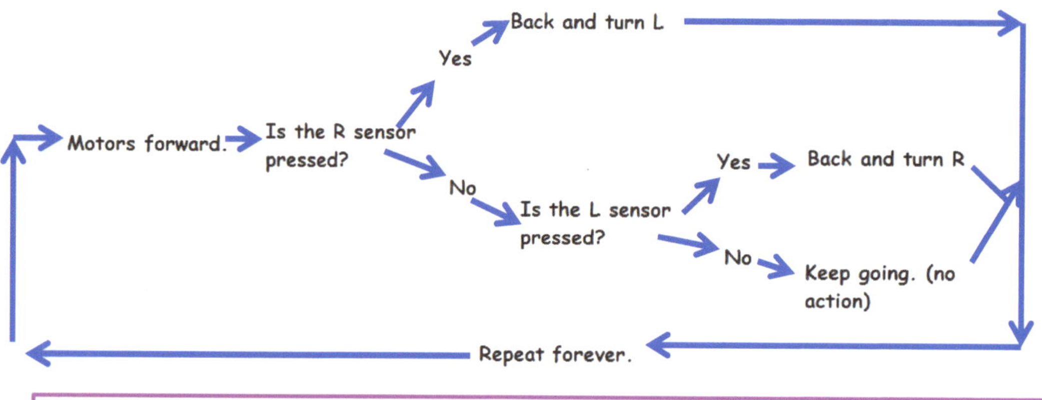

Pseudocode for this program:

```
Begin loop
        move forward, duration unlimited
                if R Touch Sensor pressed then
                        stop the motors
                        back
                        turn C quarter turn
                else check L Touch Sensor
                if L Touch Sensor pressed then
                        stop the motors
                        back
                        turn B quarter turn
                else no additional action
                end if
Loop forever
```

Intro to Data Wires

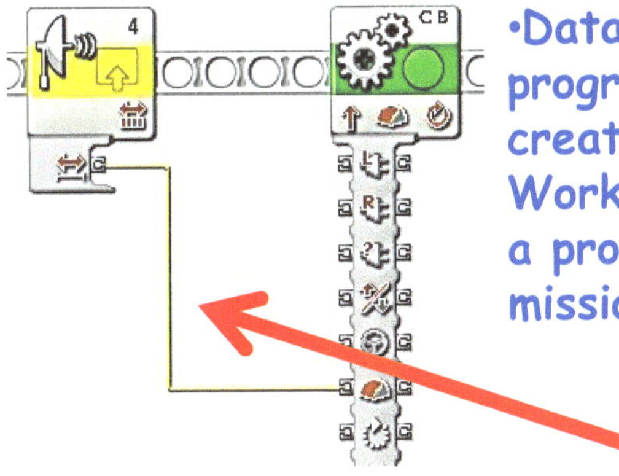

•Data wires are an important part of NXT programming, and open up an entire new aspect to create more advanced and complicated programs! Working with data wires will increase your skill as a programmer, and to complete more challenging missions!

Question: What is a data wire?

•Data is information, and just about all program blocks need this data to complete a process. You already know how to change the settings on the configuration panel of a Move block to change the duration of the motors, the power, the motors, etc. before the motors actually move. By doing this, you are putting in Input Data into the program block.

•In NXT - G, you actually have program blocks that create data for other blocks. You can have a sensor block take a reading and then control a motor by sending the data from the sensor block DIRECTLY to the Move block, WHILE it is running!!! This makes programming so much more flexible!

Data coming out of a block is called *output data*, and if it goes into another block, it is *input data.*

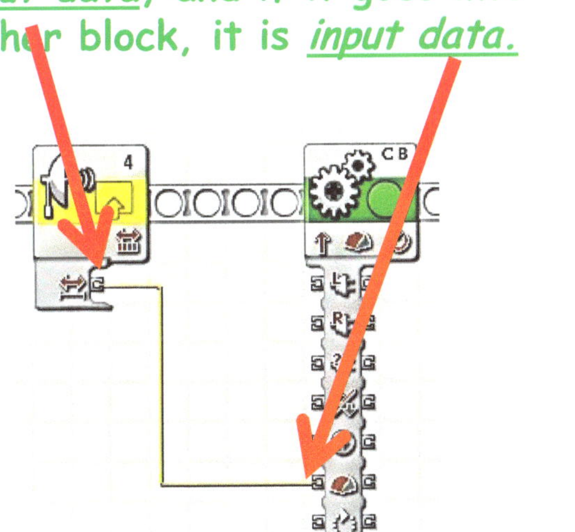

Here is how you get the data hubs to open: press on the line on the bottom of the block:

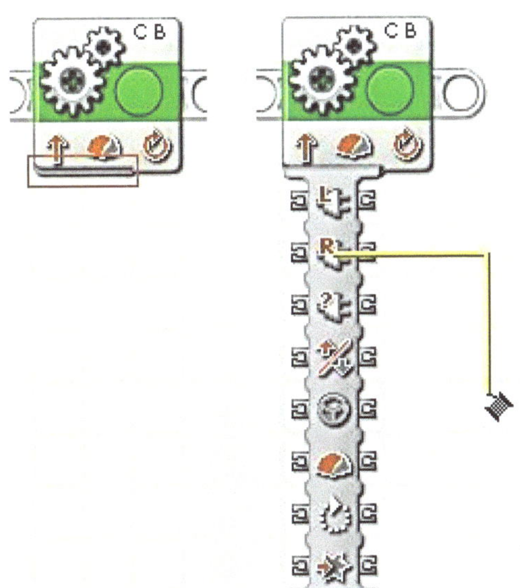

How to Use Data Wires

•Sometimes using data wires can be tricky, but with a little practice, you will be able to program some really advanced programs!

First, you need to get the data hubs to open. Here is how you get them to drop down from the programming blocks. To get the data hubs to open, press on the line on the bottom of the block:

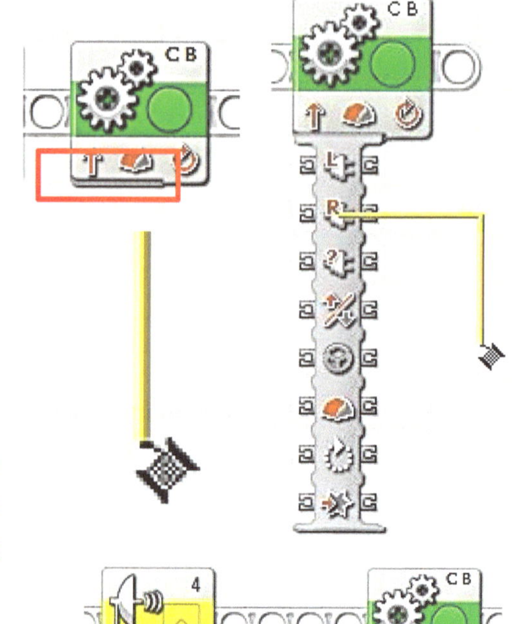

Next, you need to "unroll" the data wire from the data port. I always start FROM the RECEIVING port and go back to the OUTPUT port. It will look like a little spool of wire. You then drag this little spool with your cursor to the other block that you are transmitting data to. Make sure it attaches to a port.

If the wire is behind the data hub, you can move the data wire around on the workspace so that you can see it more clearly by pressing on it with your mouse cursor. This also helps to make the program more organized when you have multiple data wires connecting multiple blocks.

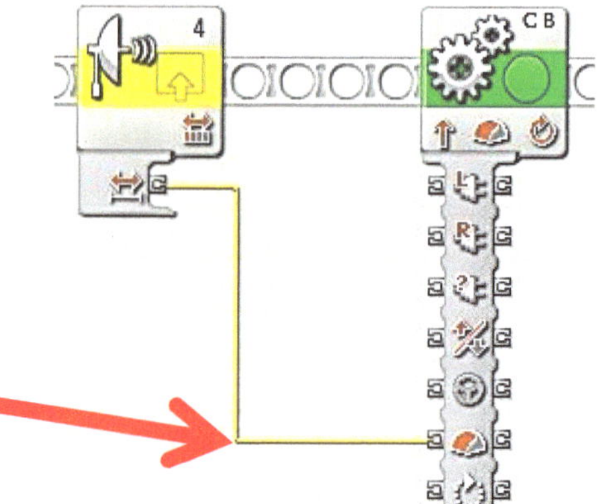

Data wires also come in different colors, because there are different types of data that can be transmitted.

Yellow data wires transmit number data (such as power levels, distances).

Green data wires transmit logic data (is the condition true/false?).

Orange data wires transmit text data (letters/words).

Gray data wires are "broken" and do not work. This happens if you attach a data wire to the wrong type of data plug. You cannot download or run a program that has a gray data wire.

Introductory Data Wire Exercise

•You are going to have the robot go forward and then decelerate to a stop before hitting a wall. How would you program this without data wires?

First of all, what is deceleration? SEE NEXT PAGE!!! Then come back!

Data Decelerate.rbt: Use an ultrasonic sensor and a Move block to gradually slow and stop the robot. The goal is for the robot to be at the same power level as the distance in cm from the wall, with the Move block being adjusted directly proportionately to the distance detected by the ultrasonic sensor, with the ultrasonic sensor controlling the Move block in real time.

Drawing of Course:

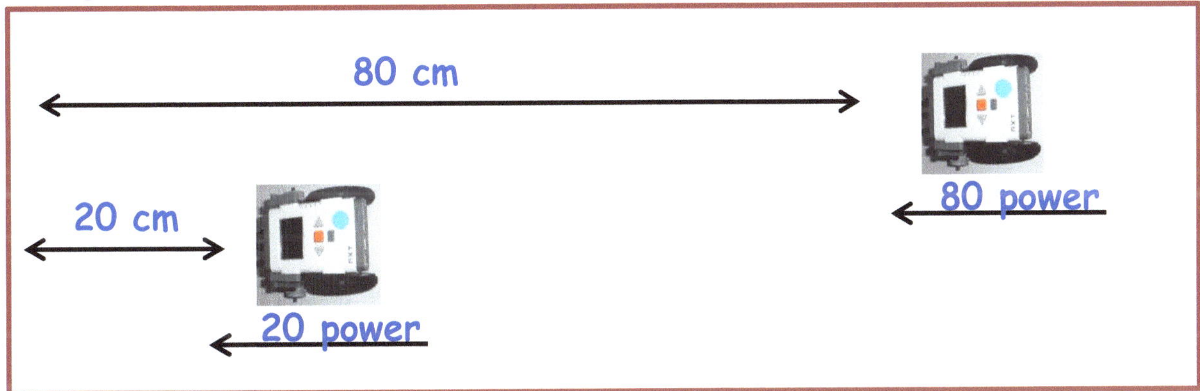

Program File Name:

Use a Sensor block, a Loop block, and a Move block. Use a data wire.

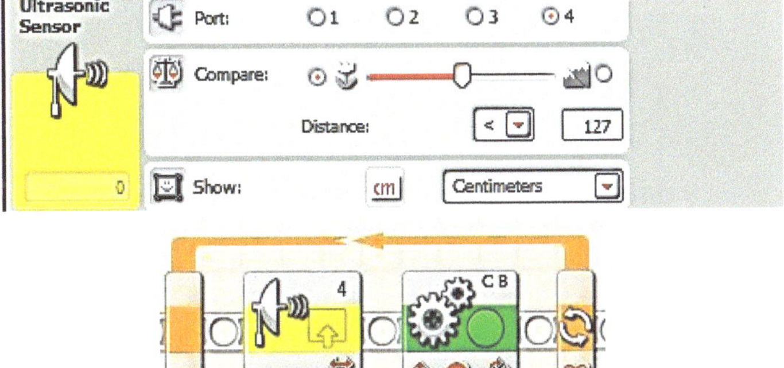

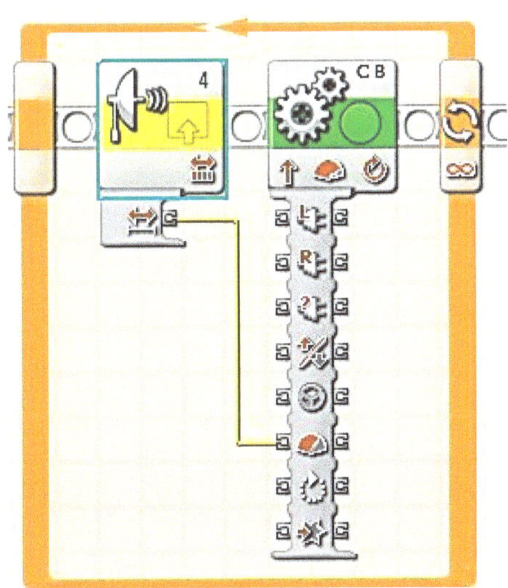

Velocity, Acceleration, and Deceleration

Velocity

Is closely related to rate.
Rate in a particular direction of an object is called velocity. It is the speed in a direction, example: 65 mph west.

Acceleration

It is a change in velocity per unit of time. Any object that is changing in rate, direction, or both is said to be either accelerating or decelerating.

Acceleration = final rate (speed) – original rate (speed) ÷ time.

Let's say you were pushing a heavy cabinet across the floor. At first, the cabinet is not moving. It is at rest and has a rate of zero! After applying a force (push) the cabinet has moved 4 feet in 2 seconds. What is the acceleration of the cabinet?

Acceleration = 4 ft/s – 0 ft/s ÷ 2 seconds
= 4 ft/s ÷ 2 seconds
= 2 ft/s/s

Deceleration (an acceleration in a negative manner) is really the same thing as acceleration where there is a decrease in speed, a slowing down. You can find it using the acceleration formula. If the object comes to a stop, the final speed is 0.

Here is a problem to try:
Mary was swimming at a rate of 8 feet per second when she started to slow down. She slowed to a stop in 2 seconds. What was her deceleration?

Acceleration = 0 ft/s – 8 ft/s ÷ 2 seconds
= -8 ft/s ÷ 2 seconds
= -4 ft/s/s

The negative sign means that this is deceleration. Mary is slowing down 4 feet per second.

Using Data Wires with Rotation Sensors

Sensor blocks (yellow stripe) create output data for other blocks. Today we are going to learn how to use Rotation Sensors to create data and transmit this data with a data wire to a sound block.

EXERCISE: Volume Control.rbt: Attach a motor to the robot and program it so that the rotation sensor tachometer controls the volume of a sound produced by the robot.

Which ports should you connect the data wires to?

____

____

Pseudocode this program after you get it to work!

Pseudocode for this program:

____

____

____

____

____

____

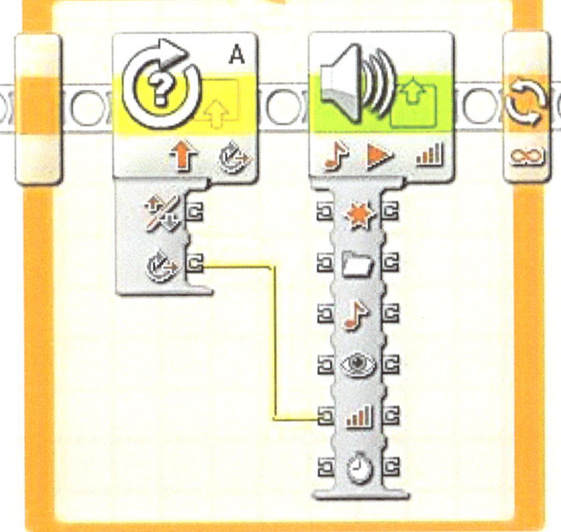

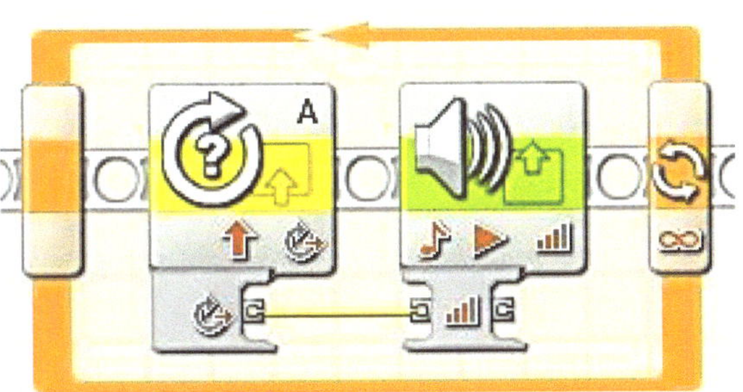

Math Blocks

•A math block is a type of data block, which means that this program block can manipulate data. In this case, what you are going to do with this data is some arithmetic operation. The data blocks are located on the complete palette.

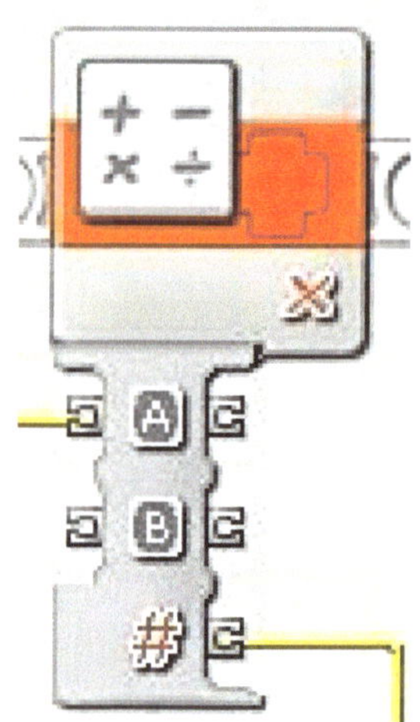

Computers are known for completing math very well, but our math blocks do have some limitations. Most calculators do have a range that they operate under, and this math block only computes numbers that are in the range of -2147483648 to 2147483647. Another problem is that this block only completes integer operations, which is fine for addition, subtraction, and multiplication. Division is another factor, we will discuss in class about this.

Here are the functions this block completes:
Addition (+) [default]
Subtraction (-)
Multiplication (x)
Division (/)
Absolute Value (|X|)
Square Root (√x)

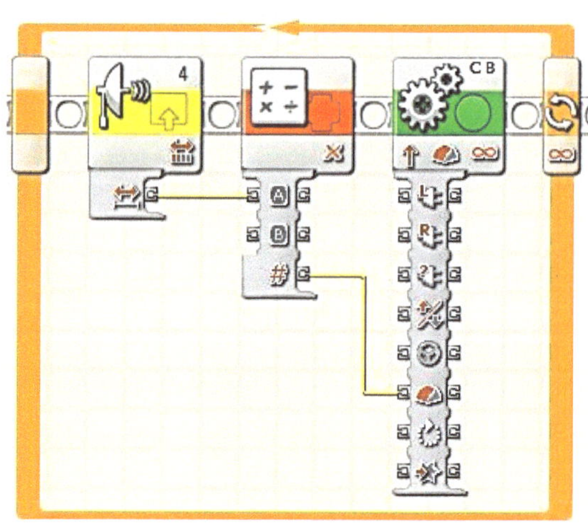

Exercise: Ultrasonic Math: In this mission, use data wires and an ultrasonic sensor to have the distance detected multiplied by 2 so that the speed is 2x the distance. Use a math block and data wires. Use a loop.

Program File Name: ______________________________

MAKE FLOWCHART:

Using Lamps

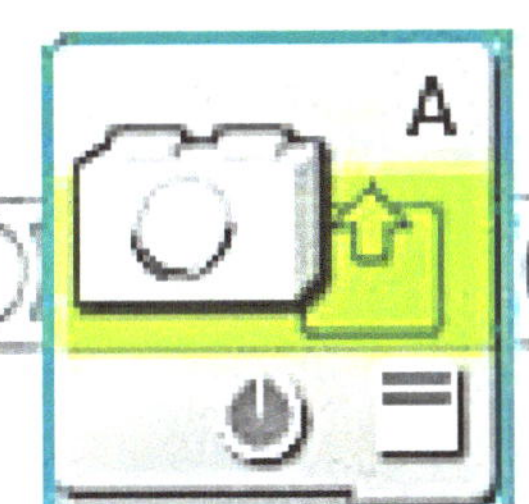

•A lamp is an actual LEGO® brick that has it's own programming block. It is an Action block. You attach the brick on to the robot and connect with the special Lamp converter cables into Port A.

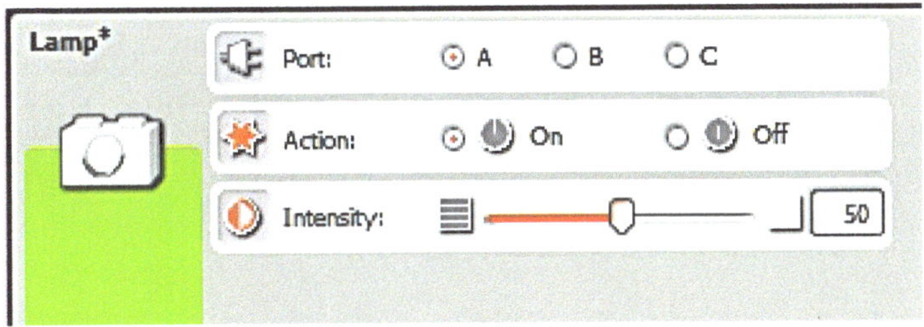

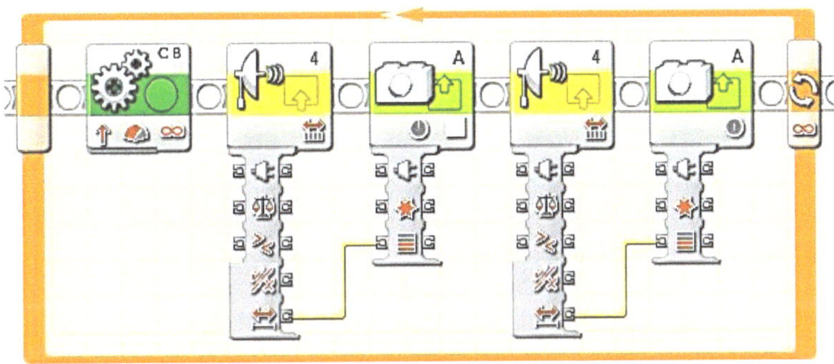

You can adjust the intensity of the Lamp. You will need one Lamp block to turn a lamp on and a 2nd Lamp block to turn it off.

Exercise: Ultrasonic Shine a Light: In this mission, use an ultrasonic sensor to have the distance detected control the Lamp. Have it so that if it is a close distance detected, then turn the Lamp on, if far, then turn it off.

Program File Name: ______________________________

MAKE FLOWCHART:

Data Wire Switches

•We are going to look at our data decelerate program and then build upon this knowledge to learn how to use a sensor block with a switch, instead of a switch set to a sensor setting.

Have the robot decelerate to a stop that actually stops. Compare this with the data decelerate program (see page 104). How is this different than our previous program?

The Switch block's value option:

So far, we have only used the sensor option of our switch blocks. You can also set it to the value option. We are going to set it on the Logic setting, which will be a True-False statement. This is going to make your programs way more useful and adjustable using the true-false (Yes-No) statements.

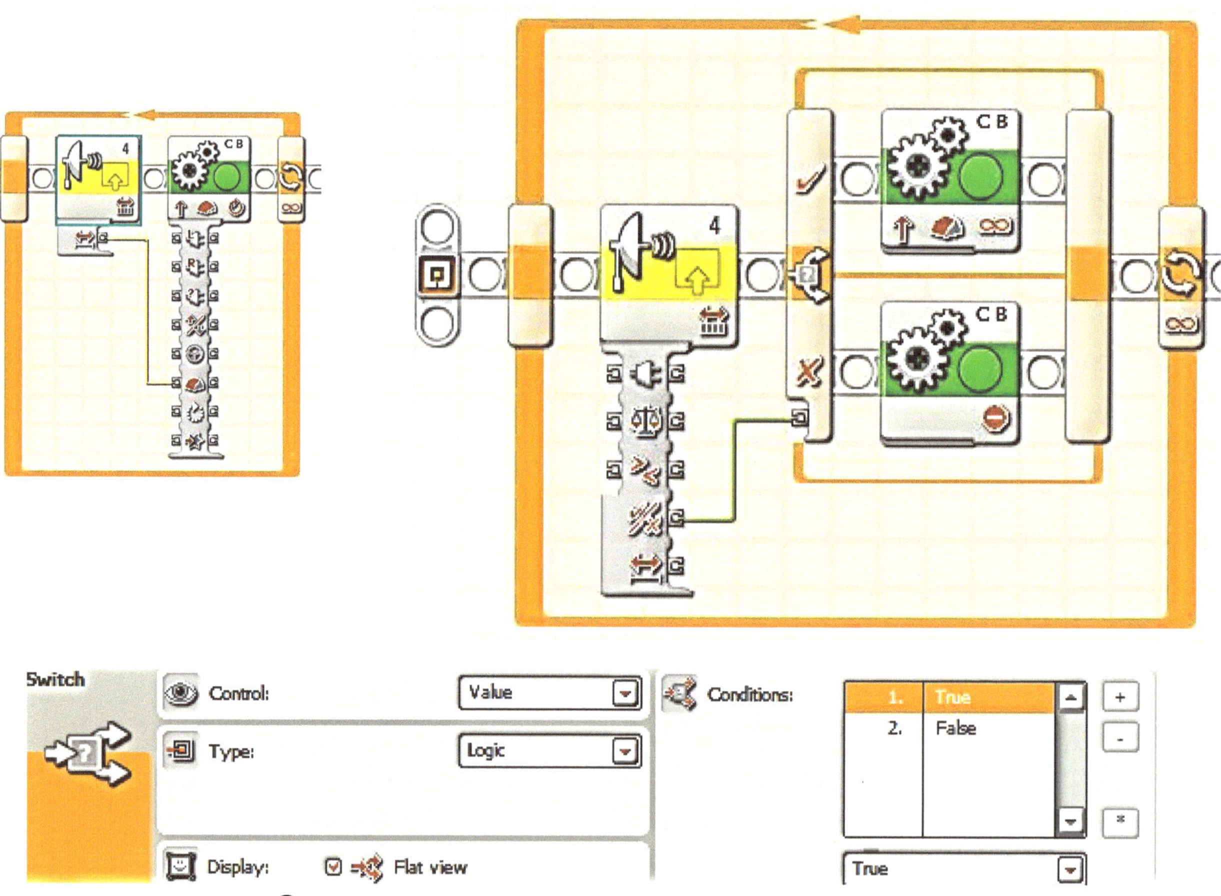

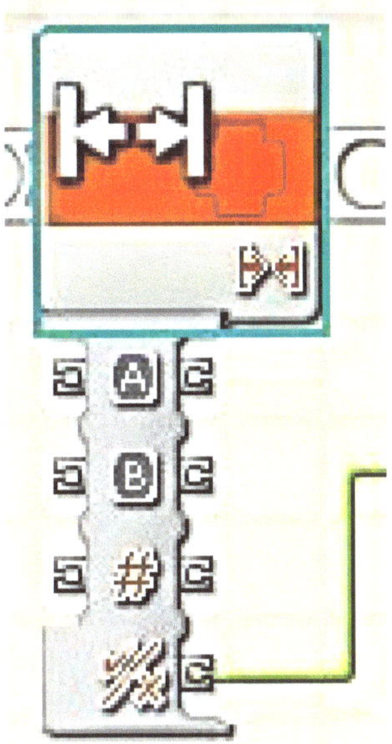

Range Blocks

The range block is a type of data block that is math related. It's purpose is to decide if a number is within or out of a range of numbers. The ranges can be set by being typed in, by using the sliders, or by the data wires. The output logic (true/false) signal will be sent by a data wire. If it is a logic wire, it will be GREEN.

Now we are going to use it so that we can have the robot detect a range of distance and then move so that it can make movements to either inside or outside of that distance. This way, we can make it follow an object (or your hand/another robot) and keep a certain distance from that object if it moves. Have the range be between 40 - 60. If the range is [40 to 60] have it move, if not, then stop.

Program file name:

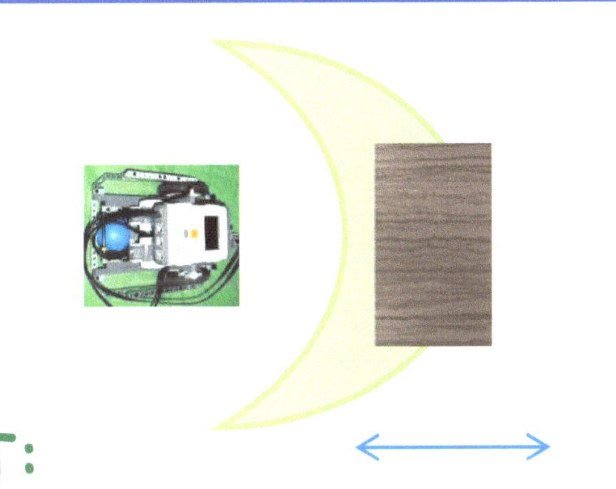

MAKE FLOWCHART:

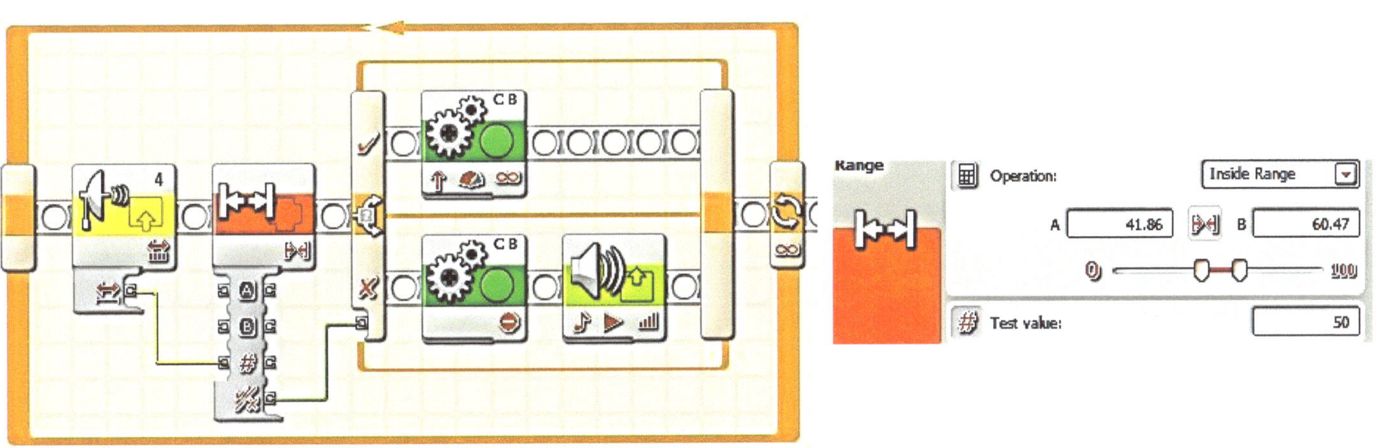

Using Bluetooth®

Programs (1)

 Bluetooth File Transfer

What is Bluetooth®?

•Bluetooth® is a wireless system that communicates data short distances, especially from a stationary device to mobile remote devices, creating personal area networks (PANs) of communication. They are seen everywhere from phones to computer mice to printers, and now between our robots and computers!

Because Bluetooth® is a technology that makes it possible to send and receive data without using cables, you are going to establish a wireless connection between your computer and your NXT™ or from one NXT™ to another NXT™. You can transfer programs between your NXT™ and PC or control one NXT™ by sending it messages from another NXT™.

The LEGO® MINDSTORMS NXT™ includes a wireless Bluetooth® node that enables the NXT™ brick to communicate with other Bluetooth® devices. For it to connect to our PCs, we have to plug a Bluetooth® dongle into the USB port on our computers.

Using Robotic Remote Control

NXT™ Remote:

•Once you connect the Bluetooth with your robot, open up NXT Vehicle Remote. Check which COMport you have and then connect this in NXT™ Remote.

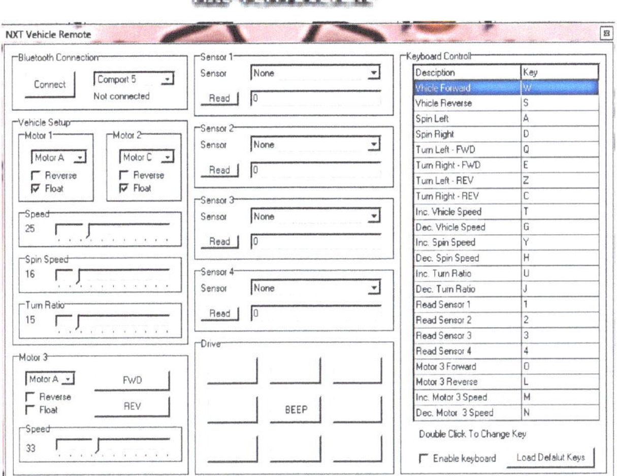

EXERCISE #1: Try out the remote using by driving around the obstacle course / or cones that your teacher sets up.

EXERCISE #2: Your mission is Operation Christmas Hershey® Round-up! You are given a specific Kisses® color, and you must build and remotely control your robot through the Grinch's cave to see how many you can gather in a specific time period, and place them in a specific area. There will be 2 rounds of this.

EXERCISE #3: Hershey® Kisses® target practice! Use the launchers and remote control to shoot at a specific color of Hershey® Kiss chocolate.

Mission Completion Information

Instructor Check: ☐

Name: ______________________

Teammates: ______________________

____

____

____

Robot/Computer: ______________

Class (es): ______________

Total Mission Points: ☐

Roboscope

Remote Robotic Search and Rescue Mission

1) Mission Description Scenario:

____

____

____

____

2) Mission Equipment / Resources:

____

____

____

____

____

3) Mission Plan: (Please List) PART 1:

- ______________________
- ______________________
- ______________________
- ______________________
- ______________________
- ______________________
- ______________________
- ______________________
- ______________________
- ______________________

4) Mission Plan: (Please List) PART 2:

- ______________________
- ______________________
- ______________________
- ______________________
- ______________________
- ______________________
- ______________________
- ______________________
- ______________________
- ______________________

5) Draw Manipulator(s):

UNIT 9

SCIENCE RELATING TO ROBOTIC PROJECTS

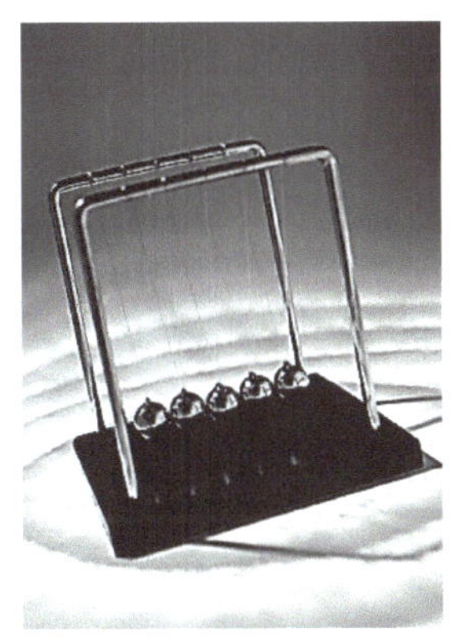

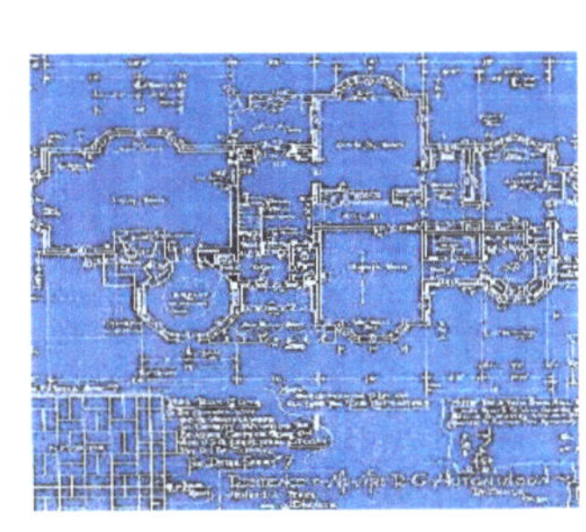

In this unit, you are going to build and apply upon your knowledge of robotics with advanced robot projects and learn all about the science behind the structures. The robot builds are more challenging, and you may need to refer to your unit on advanced programming to complete these projects and missions.

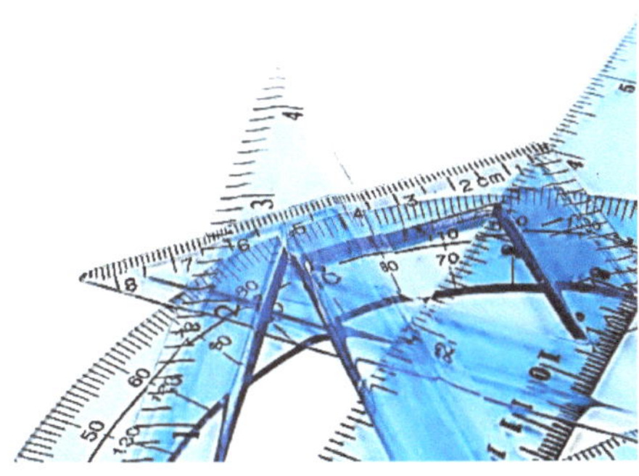

Intro to Walking Robots

We have made many types of robots that move due to wheels or treads. While wheels and treads are great, and allow for fast movements, sometimes a robot may need a different kind of locomotion ability. One way to move robots is by the use of LEGS! In this unit, we are going to study about legged, walking robots and their applications.

Legged, walking robots can do many things that wheeled robots cannot. For example, the DARPA Cheetah, Big Dog, and the LS3 robots can climb over extremely rough terrain, such as large boulders, shrubbery, deep snow, and fallen trees. Another reason to make legged, walking robots is that they are designed after walking creatures such as horses, insects, and even humans. After all, building robots that are humanoid (human-like) is a growing area in the field of robotics, and is really cool!

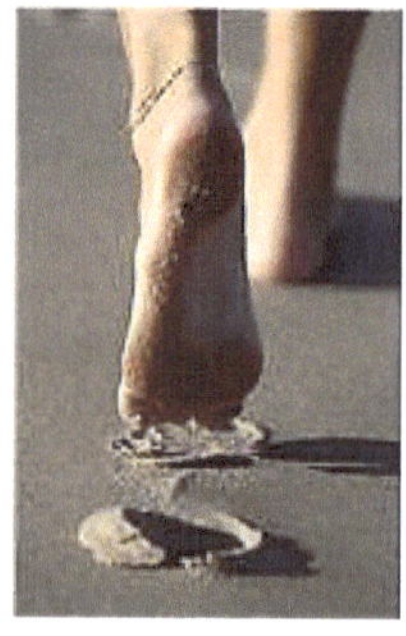

•Before we build a walking robot, we need to ask the question: what is walking? Have you ever thought about how does a creature walk exactly? Could you explain the mechanics of walking in detail to someone else?

•Walking (also called ambulation) is a gait of locomotion (movement) for creatures with legs. It is when you lift one leg at a time while supporting yourself with the other leg(s) (some creatures may have more than two legs). This is different than running, in which for an instant two legs are off the ground at the same time.

•How can you actually lift one leg at a time while supporting it with the other? It deals with a branch of physics called statics.

Statics

Statics is the study of the forces in nonmoving objects. Let's look at it in a little bit more detail and how it relates to the laws of balance:

All physical objects are made up of matter. What is matter? Matter is anything that takes up space. There are three main phases of matter: solids, liquids, and gases. (The 4th phase is plasma). Some properties of matter include color, size, shape of that matter. We are going to see how an object's matter relates to the field of statics of a solid object.

An object has mass. Mass is the amount of matter in an object. I like to say, the amount of "stuff" in an object. Let's look at a real life example: If you have two glasses, and one has water in it, which glass has more mass? The one with the water, because even though the empty glass still is full of air particles (which also have mass - but are farther apart than liquid molecules) the water is more dense and is heavier, has more "stuff" in it, and thus has more mass. This is different than weight, as weight is the measure of force on the object caused by gravity. (It is the amount of pull of gravity upon an object).

What is gravity?

Gravity is a force of attraction between two objects. The smaller object is pulled towards the object that has more mass. Here on earth, one object is the earth and the other is, well, any object on earth. It is the force that causes a book to fall to the floor when you drop it. Gravity pulls on all things all the time. The force of gravity pulls harder on things that contain more mass. When we weigh objects, we are finding out how hard gravity is pulling on them.

Center of Gravity

The Center of Gravity (CG):

What is the Cg? It is a point in an object where the rest of the object is balanced around when pulled upon by the force of gravity. Every object has a center of gravity. Find the center of gravity of a pencil or a spoon by balancing it on one finger. When you get it balanced, you have found the center of gravity of the pencil. TRY IT! You have a center of gravity too. When you stand up, your CG is in line with your body, But if you bend over at the waist, your CG moves forward, and if you lean too far forward, you may have to take a step or put out your arms so that you do not fall!

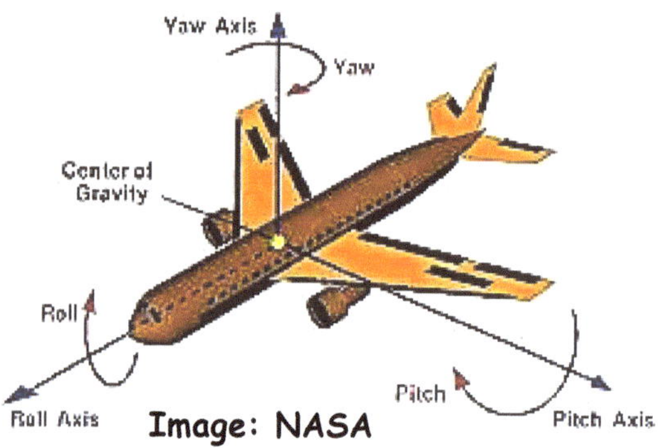

Image: NASA

Why does a lamp stand on a table without falling to the ground? Gravity is pulling the lamp down towards the earth but why is it not on the ground and why is it staying still? The lamp stays stationary due to a force that matches the force of gravity. This force is the table that holds it up. An object won't fall to the ground if something holds it up. When you have two matched forces, it is said to have a balanced force. It is at a state of equilibrium.

The table is the supporting force which holds the ball up. It matches the force of gravity.

Gravity pulling the ball down

Balanced forces

Center of Reaction

There is another center point of objects, the center of reaction (COR). The center of reaction is the center of all the supporting forces acting upon the object.

Equilibrium occurs when the center of reaction equals the center of gravity. If the COR and the CG are lined up, the object will be stable.

What about tall objects? Let's look at how statics make sure that tall skyscrapers and large slanted objects do not fall over due to equilibrium.

In statics, something is balanced if its vector CG is within its supporting base. The supporting base is what the base of the object is in straight lines.

CG ≠ COR: not stable
CG = COR: stable

Look at the pictures below of LEGO® brick structures. Look at the vectors of the forces of the CG and COR. Which structures are stable? Which structures are not stable? Why?____________________________________

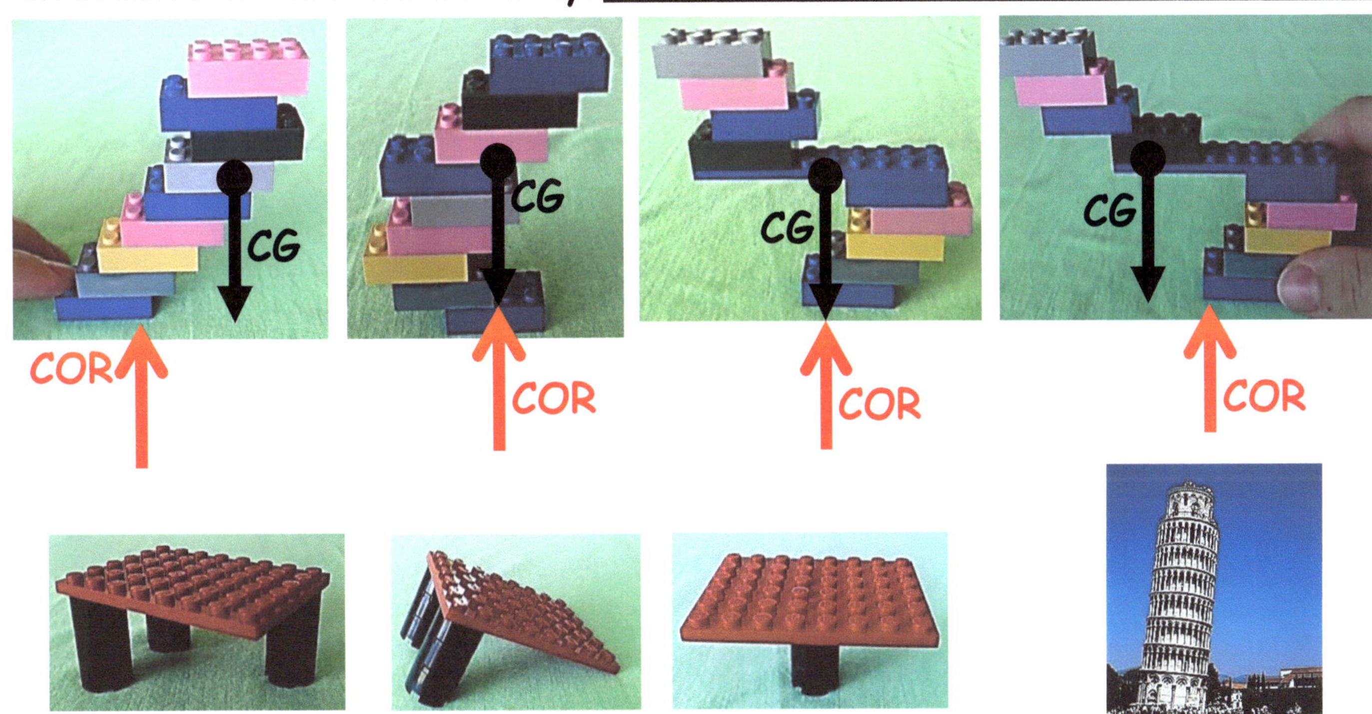

Building Robot Legs

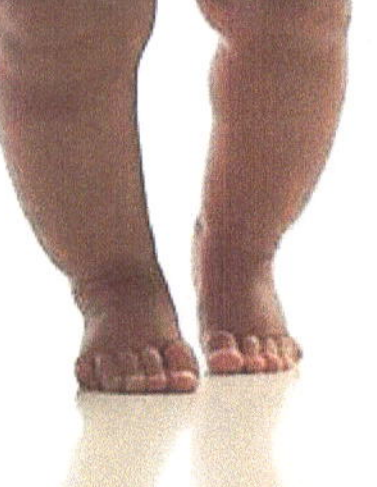

Now that we have looked at the physics behind the science of statics and stable structures, we are going to use this knowledge to engineer good legs for a walking robot.

Here is an example of a leg built out of gears. The top gear is centered on an axle which can be attached to a motor. This gear then transfers force and motion to the bottom big gear via an idler gear. Build the leg and see how it works.

Attach this motor to the IB and program it to go 7 rotations forward and then 7 rotations back.

What type of motion does this leg produce? How big is the stride?

____

____

____

Next, try to build a leg that has a fulcrum. Test this leg out as well. Compare it to the leg structure above. How is it different? Which do you think is better?

____

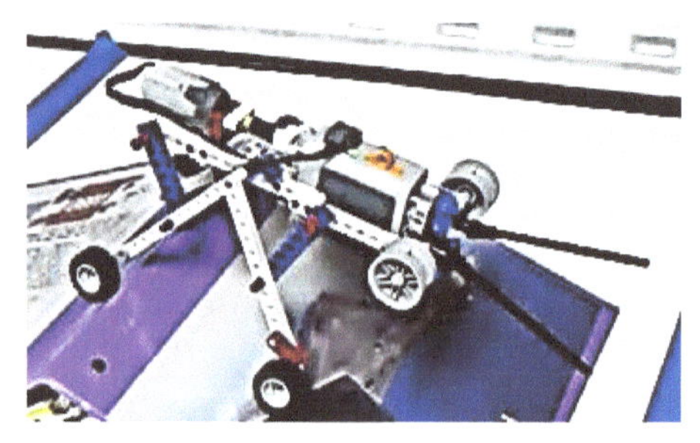

Dynamically Moving

Now that we have looked at the physics behind statically stable structures, we are going to look at another field of physics, dynamics. Dynamics has to do with matter in motion.

When something walks, we not only have to consider rules of statics, but also of dynamics, because your matter (mass) is in motion when walking or running.

When walking, you are moving, and thus you need to be in dynamic balance, which is when you have forces that oppose the forces of gravity and statics. When you walk, as a bipedal creature, you shift your center of gravity over your supporting foot for an instance while lifting and stepping forward on the other, then you repeat this again on the other foot.

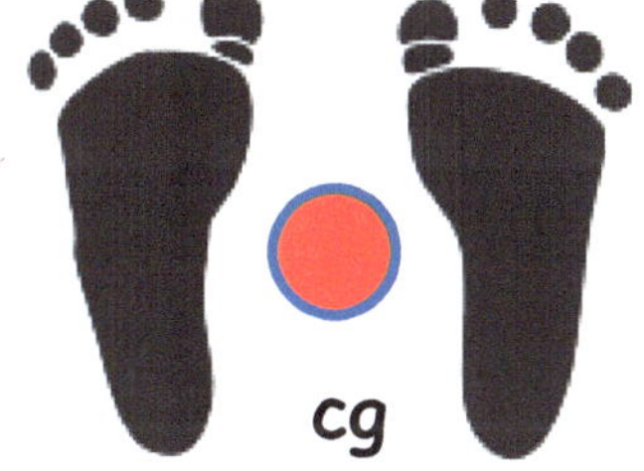

The foot that has the center of gravity over it must have its cg within its supporting base, but then for an instance, it shifts out and over to the other foot. You can do this without falling down! This is because you are statically balanced and dynamically balanced!

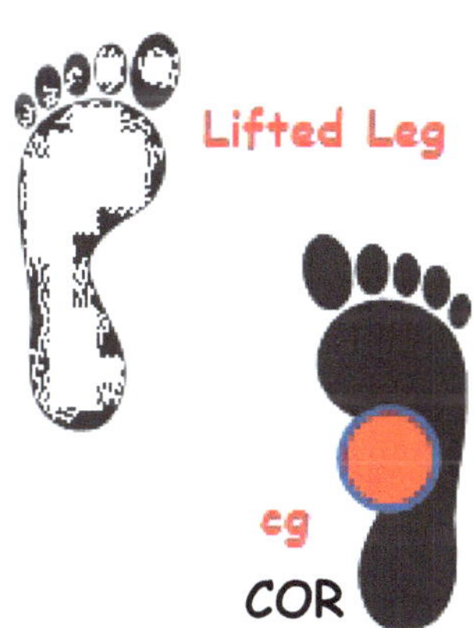

As you change your gait to running, it is more important to be dynamically balanced than statically, because you are moving at a really fast rate and you can cover more ground in a stride. Actually, if you were able to freeze in the middle of a running stride, you would not be statically balanced, and would topple over!

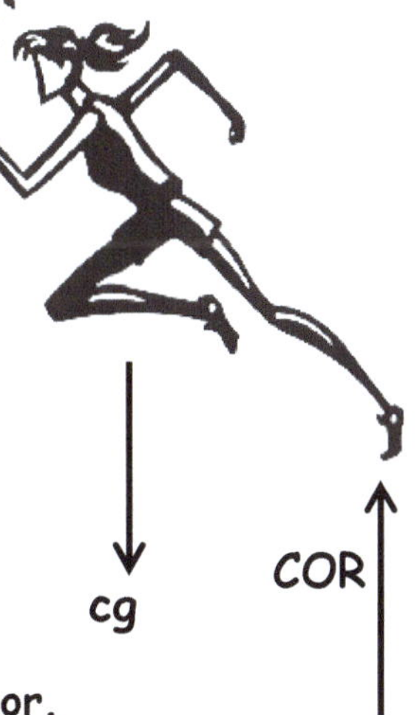

Biped Robot

Here is a neat robot build that demonstrates statics and dynamics: The Turkey Walker!

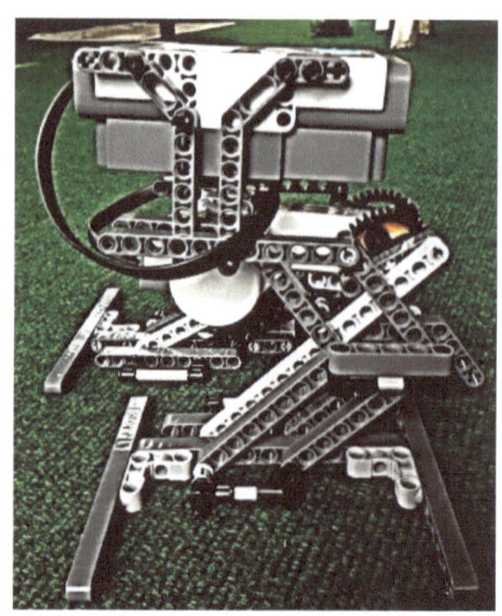

Build the model according to the building directions. Then write a simple program that tests out the legs of the Turkey Walker.

The Turkey Walker is a biped robot that has large feet with a series of levers. It makes a walking movement due to its shifting of its center of gravity back and forth from leg to leg.

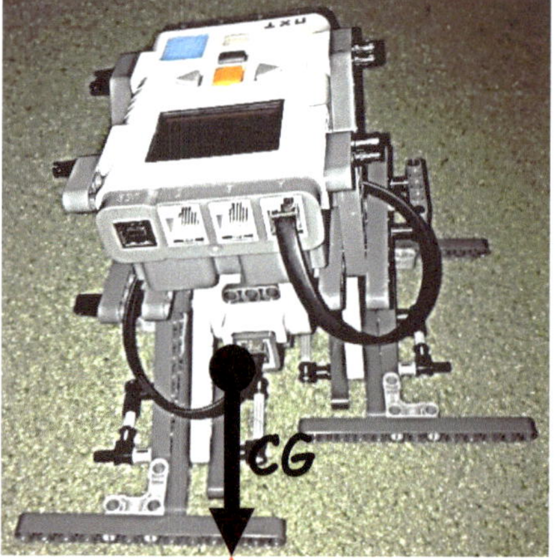

Can the Turkey Walker go backwards as well? How fast/slow can it go? What kind of missions can this robot complete? Is it useful?

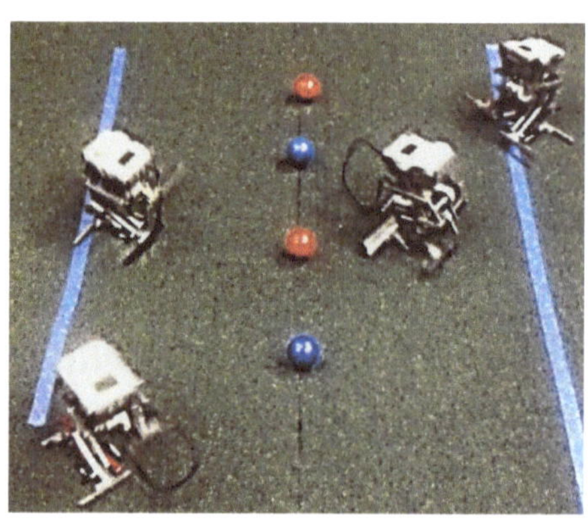

With your class, play a modified ball game using the Remote.

Robotic Arm

As the field of robotics is growing, robot engineers are making great strides in the field of robotic arms. Robotic arms are being used in many different areas and are changing the fields of industry and medicine.

Industrial robotic arms make productivity better, safer, and faster, allowing for the production of higher quality products at a cheaper price for consumers. Space robonauts usually have robotic arms to do tasks in the unfriendly environment of space.

Prosthetic arms give amputees the opportunity of being able to have the freedom of using tools, opening doors, and giving a hi-five.

Other robotic arms can be used as picker uppers, pancake flippers, or even back scratchers!

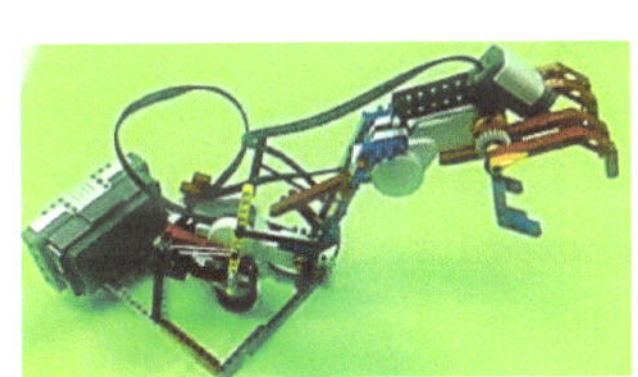

Our robotic arm uses servo motors to allow it to twist left and right, lift up and down, and open and close the hand/claw. These movements are called the degrees of freedom.

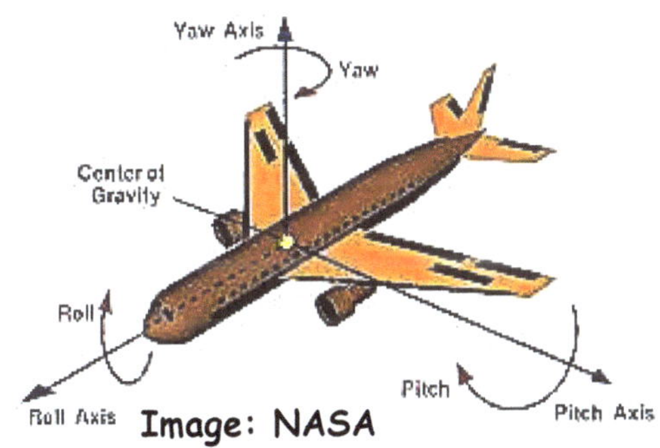

Image: NASA

Degrees of Freedom (DoF)

Degrees of freedom (DoF) are the movements of an object in an up-and-down, left-and-right, and forward-and-backward manner. Each DoF has a specific name:

1) Pitch is up-and-down movement
2) Yaw is left-and-right movement
3) Roll is a rotation around such as using a screwdriver

Photo: NASA

DoF of humans:

- Human shoulders can yaw, pitch, and roll (3DoF), human elbows can only move up and down (only 1DoF), human wrists can yaw, pitch, and roll (3DoF)
- So adding them up, the entire human arm has 7DoF

How many DoF will your robotic arm have, and how many motors will you need to use for each DoF?

Build and Test the Robotic Arm

- Find out which picking up sequence is the most efficient for picking and placing objects or a piece of crumpled paper.

- Your sequence must start in the resting position, use all six movements at least once, and then return to the resting position for a total of 8 strokes.

How could you program this? Make a flowchart:

This robotic arm is equipped with sensors. How are you going to program these sensors?

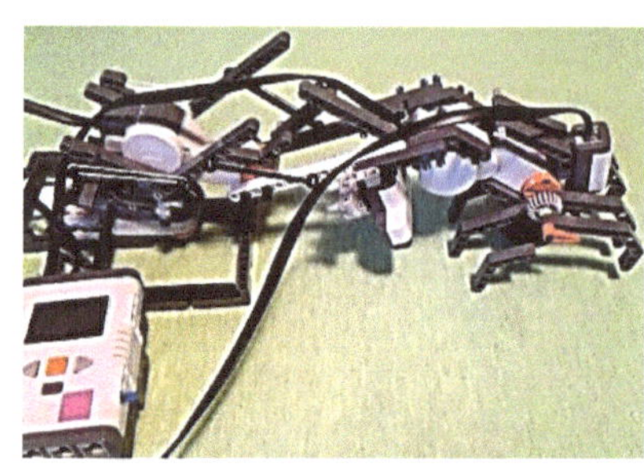

Extension activity: Make this arm attach to a vehicular robot or make it attach on your arm! What do you have to do to modify this robot?________________________________

____

____

____

Photo: NASA

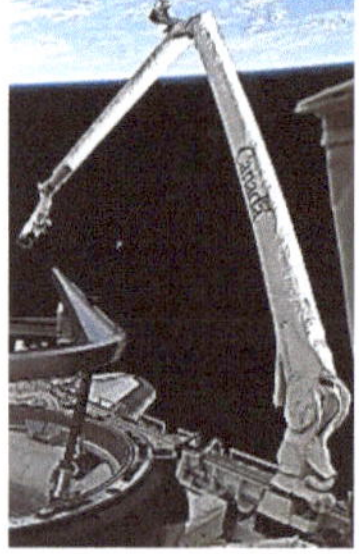

Photo: NASA

The Shuttle Remote Manipulator System (SRMS), or Canadarm, is a mechanical arm used on the Space Shuttle to maneuver a payload from the payload bay of the orbiter to its deployment position and then release it.

Quadruped Robot

Now we are going to apply our knowledge of legs and statics and dynamics to build a quadruped robot! Please meet Quadimoto!

Quadruped means 4 legs. Building and programming 4 legs can be a challenge, but we are going to create these legs to attach two legs per motor (motors B and C).

Build the model according to the building directions. Then write a simple program that tests out the legs of quadruped robot.

---

4 - Legs Walking

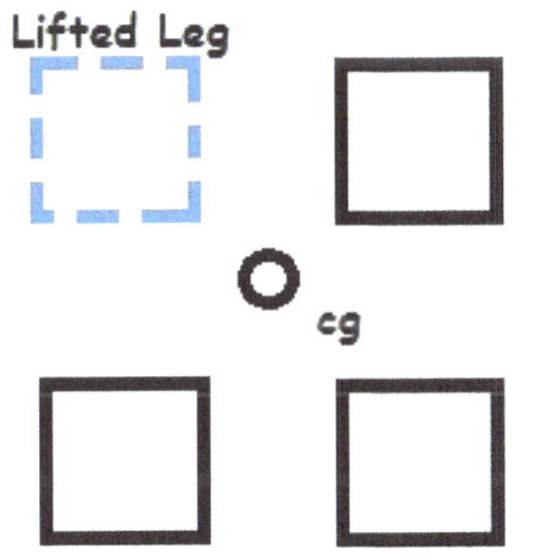

Draw the outline of the supporting base area on the diagram. Does the cg fall inside of the supporting base (where the center of reaction is located?) Is this walking robot stable?

Side view of walking strides

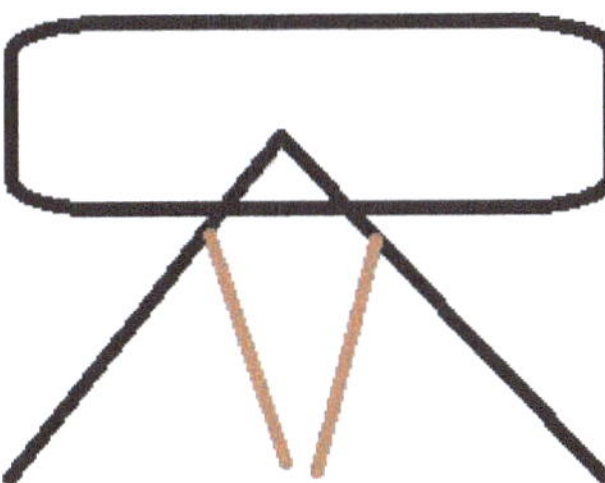

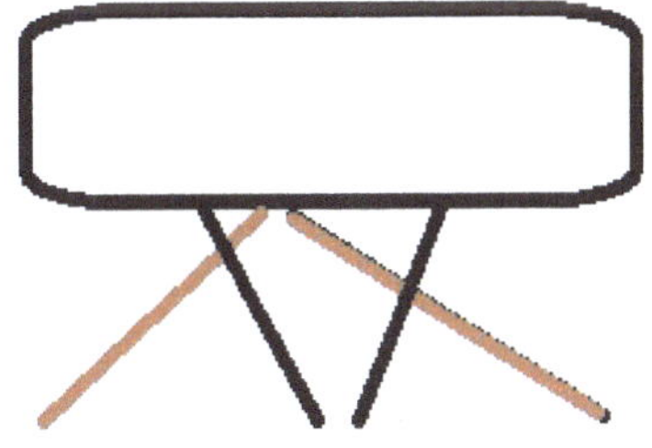

4 Legs Running

Adjust the legs settings on the elbow Technic™ beam attached to the torque providing axle from the motors. Make the settings of the legs so that the robot has a higher step, and are spread out.

When a creature is running, it has at least more than 1 leg off the ground at the same instant. Look at the diagram.

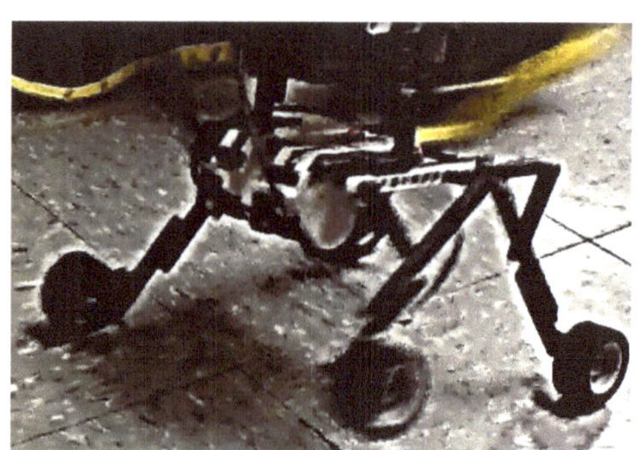

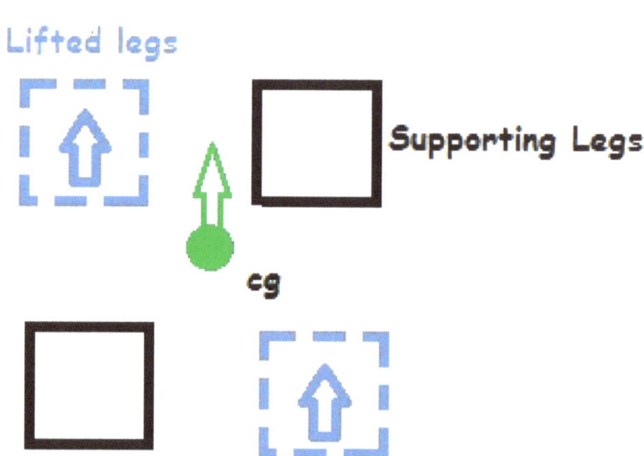

Again, draw the outline of the supporting base area on the diagram. How does the cg fall inside of the supporting base (where the center of reaction is located) if the mass of the robot is moving? Is this running robot stable statically and dynamically?

Side view of running strides:

Can this robot climb over objects and rough terrain? Can it turn? Test it!

____

____

____

____

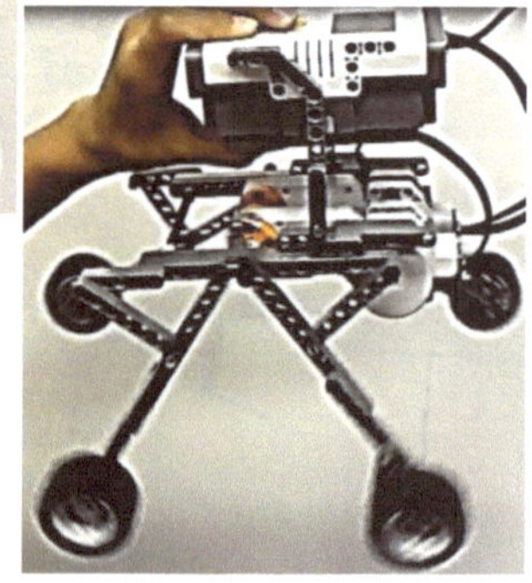

Robotic Automated Independent Exploration Mission (RAIE)

This is our year-end, class mission challenge that will extend for two classes.

In this RAIE mission, you and your partner will work together to have your robot complete an obstacle course.

Mission Requirements:
You will start at base, and then check in at the target stations in the proper order to the finish.

Target stations include:

- Grab the loop
- Deliver the LEGO® ball
- Bump into a target to knock it down
- Shoot the target

You must use at least two sensors!

Before starting, you must come up with your plan of attack and order of operations!

Remember the 3 C's!

You must be mindful of the time you have to complete this mission! Do not waste time just trying things out and hoping that they work. I will need to see a written plan before you start on this mission, and you must estimate how much time you need for each part of this mission.

RAIE Mission Plans

Step 1:	Estimated time to complete step 1:

You need to plan as a team a robot build that will work accurately, make precise turns, and can run on carpet or wood. Skids or casters might be a good place to start.

You need to build this bot quickly, while making it strong and secure. Trusses and long connector pegs are recommended. See Class 4 Page 5 for ideas.

Keep in mind that you may need to rebuild this next class!!!!!!!!!!!!!!!!!!!!!!!!!!!!!!!!!!!!!!!

Step 2:	Estimated time to complete step 2:

Plan a path and arrange in order the target obstacles from start to finish and place them in correct order.

1. ____________________
2. ____________________
3. ____________________
4. ____________________
5. ____________________
6. ____________________
7. ____________________
8. ____________________

RAIE Mission Plans

Step 3:	Estimated time to complete step 3:

You need look at the obstacle challenge sequence and then plan out your manipulators. Remember, you have only 3 motor ports, and usually 2 are used for locomotion purposes. Are you going to use active manipulators (motorized manipulators) or passive manipulators (lacking motor)? Can you use one manipulator for more than one target obstacle?

Target Obstacle	Manipulator Idea

Step 4:	Estimated time to complete step 4:

How could you program this? Make a flowchart:

Don't forget your sensors!!!!!!

START

Robotic Review

WOW – we have come such a long way this year in Robotics Engineering class!! Everyone did such a great job working together as a team, being creative, and staying focused on the tasks!

Let's look back and see what we have accomplished so far this year. As Julie Andrews said… "♫ Let's start at the very beginning.♪"

We first looked at what a robot is: an automatically guided machine which is able to do tasks on its own, almost always due to electronically-programmed instructions.

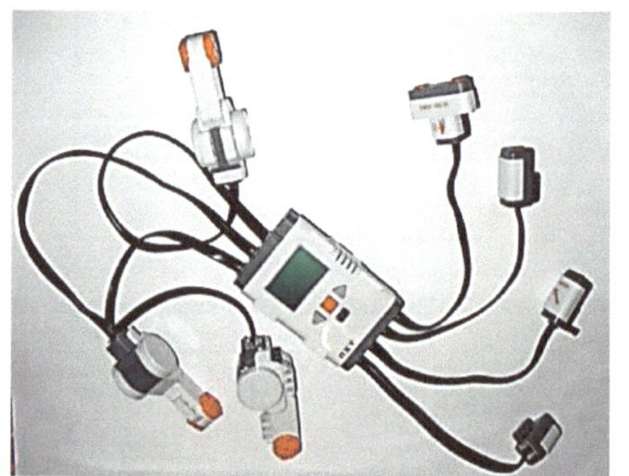

We then learned about certain features of our Mindstorms™ LEGO® robot, and all the sensors and applications.

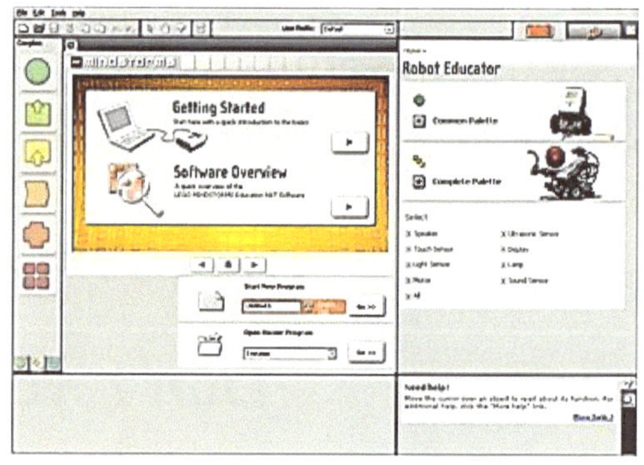

We then learned about programming, and the NXT – G programming language, and all the adjustments we can make with the programming blocks.

We also looked at the 3 C's: What are they? ☺

Then we started learning about missions. Remember that there are 3 major types of missions:1. strike mission – you go out in mission field and bump/strike a target, 2. delivery – you take an object from base to a target and leave it there, and 3. gathering – you go out to the mission field and bring an object back to base.

Then we learned about robot locomotion and how to get the robot to make movements precisely and efficiently.

Robotic Review Continued

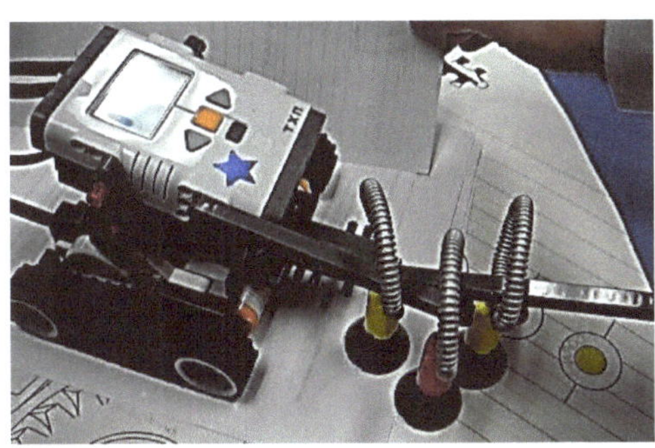

• In Unit 5, we learned how a manipulator is a device used by the robot to manipulate materials. We also studied and built pushing manipulators, hook/fork manipulators, CG manipulators, lever manipulators, collecting manipulators, dumping manipulators, and motorized manipulators such as grabbers.

Then, by completing the sensor unit, we learned how a robot must be able to react to its environment by sensing. Just like we have our 5 senses for our human body to react and gain information from our environment, our robots use sensors to acquire this information.

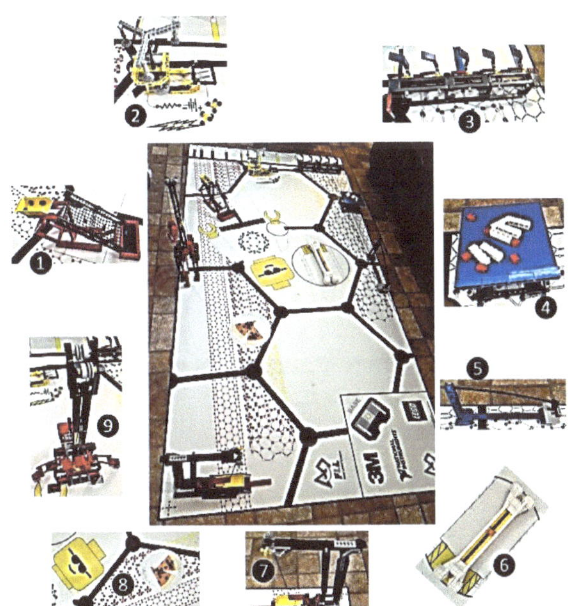

After learning about basic robotic task missions, we went on to the Nanotech Mission Challenge, exploring the world of nanotechnology and making scientific robotic breakthroughs on a nanoscale. Everyone worked hard on this mission challenge with 9 mission tasks, and I saw lots of creative and streamlined mission runs, with efficient and practical builds for manipulators.

Which one of the Nanotech missions was your favorite?

Robotic Review Continued

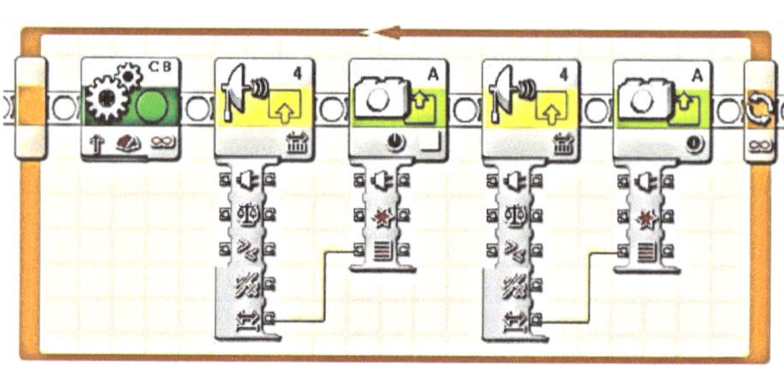

Then we did in-depth, advanced programming. The programming exercises taught you how to program the robot so that it is more autonomous, which is the ultimate goal of robotics!

In the last unit on science relating to robotic projects, you built and applied upon your knowledge of robotics with advanced robot projects to learn all about the science behind the structures, learning all about statics and dynamics, and all about robot with legs! You then finished up the year with our RAIE mission, which was a complex team challenge!

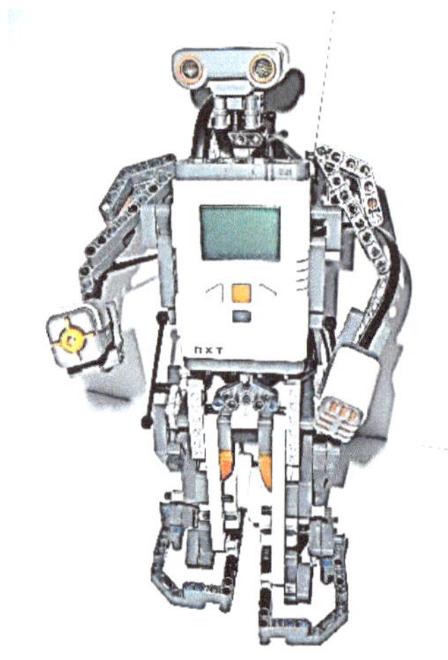

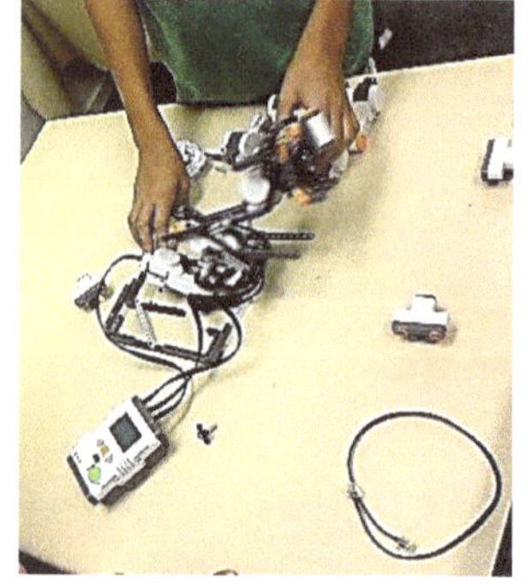

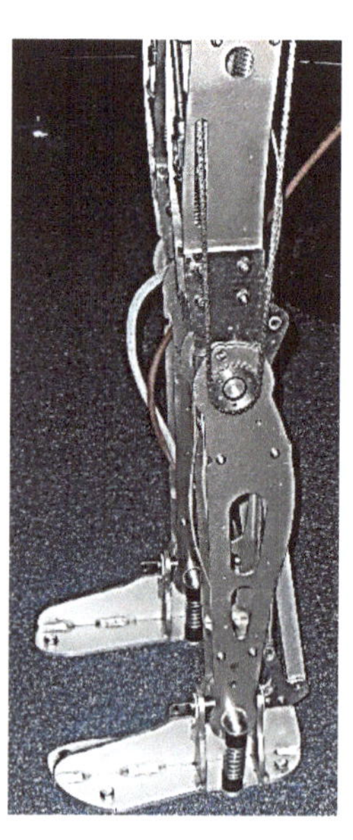

Robotics is an exciting field with applications in many parts of everyday life. Every day this field of science and technology is rapidly changing. Can you think of a need where a robot could be used to help others? Would you be interested in possibly one day going into a career in robotics?

www.ingramcontent.com/pod-product-compliance
Lightning Source LLC
Chambersburg PA
CBHW050719180526
45159CB00003B/1073

9781508428930